Introduction to Electric Circuits

To the memory of my mother and father with grateful thanks

Essential Electronics Series

Introduction to Electric Circuits

Eur Ing R G Powell
Principal Lecturer
Department of Electrical and Electronic Engineering
Nottingham Trent University

Newnes

Newnes
An imprint of Butterworth-Heinemann
Linacre House, Jordan Hill, Oxford OX2 8DP
225 Wildwood Avenue, Woburn MA 01801-2041
A division of Reed Educational and Professional Publishing Ltd

A member of the Reed Elsevier plc group

OXFORD AUCKLAND BOSTON
JOHANNESBURG MELBOURNE NEW DELHI

First published 1995
Transferred to digital printing 2001

© Ray Powell 1995

All rights reserved. No part of this publication may be reproduced in any
material form (including photocopying or storing in any medium by
electronic means and whether or not transiently or incidentally to some
other use of this publication) without the written permission of the copyright
holder except in accordance with the provisions of the Copyright, Designs and
Patents Act 1988 or under the terms of a licence issued by the Copyright Licensing
Agency Ltd, 90 Tottenham Court Road, London, England, W1P 0LP. Applications
for the copyright holder's written permission to reproduce any part of this publication
should be addressed to the publisher

British Library Cataloguing in Publication Data
A catalogue record for this book is available from the British Library

Library of Congress Cataloguing in Publication Data
A catalogue record for this book is available from the Library of Congress

ISBN 0 340 63198 8

For information on all Newnes publications
visit our website at www.newnespress.com

PLANT A TREE
FOR EVERY TITLE THAT WE PUBLISH, BUTTERWORTH-HEINEMANN
WILL PAY FOR BTCV TO PLANT AND CARE FOR A TREE.

Series Preface

In recent years there have been many changes in the structure of undergraduate courses in engineering and the process is continuing. With the advent of modularization, semesterization and the move towards student-centred learning as class contact time is reduced, students and teachers alike are having to adjust to new methods of learning and teaching.

Essential Electronics is a series of textbooks intended for use by students on degree and diploma level courses in electrical and electronic engineering and related courses such as manufacturing, mechanical, civil and general engineering. Each text is complete in itself and is complementary to other books in the series.

A feature of these books is the acknowledgement of the new culture outlined above and of the fact that students entering higher education are now, through no fault of their own, less well equipped in mathematics and physics than students of ten or even five years ago. With numerous worked examples throughout, and further problems with answers at the end of each chapter, the texts are ideal for directed and independent learning.

The early books in the series cover topics normally found in the first and second year curricula and assume virtually no previous knowledge, with mathematics being kept to a minimum. Later ones are intended for study at final year level.

The authors are all highly qualified chartered engineers with wide experience in higher education and in industry.

R G Powell
Jan 1995
Nottingham Trent University

Contents

Preface xi
Acknowledgements xiii

Chapter 1 Units and dimensions
1.1 Introduction 1
1.2 The *Système International d'Unités* 1
1.3 Dimensional analysis 4
1.4 Multiples and submultiples of units 6
1.5 Self-assessment test 8
1.6 Problems 8

Chapter 2 Electric circuit elements
2.1 Electricity 10
2.2 Electric circuits 11
2.3 Circuit elements 11
2.4 Lumped parameters 34
2.5 Energy stored in circuit elements 35
2.6 Power dissipated in circuit elements 36
2.7 Self-assessment test 37
2.8 Problems 38

Chapter 3 DC circuit analysis
3.1 Introduction 40
3.2 Definition of terms 40
3.3 Kirchhoff's current law 42
3.4 Kirchhoff's voltage law 42
3.5 The Principle of Superposition 45
3.6 Thevenin's theorem 48
3.7 Norton's theorem 52
3.8 The Maximum Power Transfer Theorem 54
3.9 Delta-star transformation 56
3.10 Star-delta transformation 59
3.11 Self-assessment test 61
3.12 Problems 62

Chapter 4 Single-phase a.c. circuits
4.1 Alternating quantities 66

4.2	Single-phase a.c. circuits in the steady state	72
4.3	Series a.c. circuits	77
4.4	Complex notation	82
4.5	Parallel a.c. circuits	91
4.6	Series–parallel a.c. circuits	95
4.7	Power in single-phase a.c. circuits	97
4.8	Self-assessment test	103
4.9	Problems	105

Chapter 5 Three-phase a.c. circuits

5.1	Introduction	107
5.2	Generation of three-phase voltage	107
5.3	Phase sequence	108
5.4	Balanced three-phase systems	109
5.5	Power in balanced three-phase circuits	115
5.6	Self-assessment test	120
5.7	Problems	122

Chapter 6 Resonance

6.1	Series resonance	123
6.2	Parallel resonance	133
6.3	Self-assessment test	137
6.4	Problems	138

Chapter 7 Nodal and mesh analysis

7.1	Introduction	141
7.2	Matrices	141
7.3	Nodal voltage analysis	147
7.4	Mesh current analysis	158
7.5	Self-assessment test	168
7.6	Problems	168

Chapter 8 Transient analysis

8.1	Introduction	172
8.2	Circuits containing resistance and inductance	172
8.3	Circuits containing resistance and capacitance	181
8.4	The Laplace transform	192
8.5	Self-assessment test	201
8.6	Problems	202

Chapter 9 Two-port networks

9.1	Introduction	205
9.2	The impedance or z-parameters	205
9.3	The admittance or y-parameters	208

9.4	The hybrid or *h*-parameters	210
9.5	The inverse hybrid or *g*-parameters	211
9.6	The transmission or *ABCD*-parameters	213
9.7	The inverse transmission parameters	214
9.8	Cascaded two-port networks	219
9.9	Characteristic impedance (Z_0)	225
9.10	Image impedances	226
9.11	Insertion loss	227
9.12	Propagation coefficient (γ)	229
9.13	Self-assessment test	230
9.14	Problems	231

Chapter 10   Duals and analogues

10.1	Duals of circuit elements	233
10.2	Dual circuits	234
10.3	Analogues	238
10.4	Self-assessment test	241

Answers to self-assessment tests and problems 242

Index 246

Preface

This book covers the material normally found in first and second year syllabuses on the topic of electric circuits. It is intended for use by degree and diploma students in electrical and electronic engineering and in the associated areas of integrated, manufacturing and mechanical engineering.

The two most important areas of study for all electrical and electronic engineering students are those of circuit theory and electromagnetic field theory. These lay the foundation for the understanding of the rest of the subjects which make up a coherent course and they are intimately related. Texts on one of them invariably and inevitably have references to the other. In Chapter 2 of this book the ingredients of electric circuits are introduced and the circuit elements having properties called capacitance and inductance are associated with electric and magnetic fields respectively. Faraday's law is important in the concept of mutual inductance and its effects. Reference is made, therefore, to electromagnetic field theory on a need to know basis, some formulae being presented without proof.

The level of mathematics required here has been kept to a realistic minimum. Some facility with algebra (transposition of formulae) and knowledge of basic trigonometry and elementary differentiation and integration is assumed. I have included well over a hundred worked examples within the text and a similar number of problems with answers. At the end of each chapter there is a series of self assessment test questions.

Ray Powell
Nottingham, November 1994

Acknowledgements

I am most grateful to a number of anonymous reviewers for their constructive criticism and suggestions. I am also indebted to the many authors whose books I have consulted over the years. My thanks are due to Eur Ing Professor Peter Holmes for encouraging me to write this book, to Allan Waters for permission to use some of his problems, to the many students with whom I have had the pleasure of working and whose questions have helped form this book and not least to my wife Janice for her patience in the face of deadline-induced irritability.

1 Units and dimensions

1.1 INTRODUCTION

In electrical and electronic engineering, as in all branches of science and engineering, measurement is fundamentally important and two interconnected concepts are involved. First we need to know what it is that we wish to measure, and this is called a quantity. It may be a force or a current or a length (of a line say). The quantity must then be given a unit which indicates its magnitude, that is, it gives a measure of how strong the force is or how big the current is or how long the line is. In any system of units a certain number of physical quantities are arbitrarily chosen as the basic units and all other units are derived from these.

1.2 The *SYSTÈME INTERNATIONAL D'UNITÉS*

This system of units, abbreviated to 'the SI', is now in general use and in this system seven basic quantities, called dimensions, are selected. These are mass, length, time, electric current, thermodynamic temperature, luminous intensity and amount of substance, the first four of which are of particular importance to us in this book. In addition to these seven basic quantities there are two supplementary ones, namely plane angle and solid angle. All of these are shown, together with their unit names, in Table 1.1. These units are defined as follows:

kilogram (kg): the mass of an actual piece of metal (platinum–iridium) kept under controlled conditions at the international bureau of weights and measures in Paris

metre (m): the length equal to 1 650 763.73 wavelengths *in vacuo* of the radiation corresponding to the transition between the levels $2p_{10}$ and $5d_5$ of the krypton-86 atom

second (s): the duration of 9 192 631 770 periods of the radiation corresponding to the transition between the two hyperfine levels of the ground state of the caesium-133 atom

ampere (A): that constant current which, when maintained in each of two

2 *Units and dimensions*

Table 1.1

Quantity	Unit	Unit abbreviation
Mass	kilogram	kg
Length	metre	m
Time	second	s
Electric current	ampere	A
Thermodynamic temperature	kelvin	K
Luminous intensity	candela	cd
Amount of substance	mole	mol
Plane angle	radian	rad
Solid angle	steradian	sr

infinitely long parallel conductors of negligible cross-sectional area separated by a distance of 1 m in a vacuum, produces a mutual force between them of 2×10^{-7} N per metre length

kelvin (K): the fraction 1/273.16 of the thermodynamic temperature of the triple point of water

candela (cd): the luminous intensity, in the perpendicular direction, of a surface of area $1/600\,000$ m^2 of a black body at the temperature of freezing platinum under a pressure of 101 325 Pa

mole (mol): the amount of substance of a system which contains as many specified elementary particles (i.e. electrons, atoms, etc.) as there are atoms in 0.012 kg of carbon-12

radian (rad): the plane angle between two radii of a circle which cut off on the circumference an arc equal to the radius

steradian (sr): that solid angle which, having its vertex at the centre of a sphere, cuts off an area of the surface of the sphere equal to that of a square with sides equal to the radius.

The scale temperature (degree Celsius) is the thermodynamic temperature minus 273.16, so that 0 °C corresponds to 273.16 K and 0 K corresponds to -273.16 °C. Note that we write 0 K and 273 K, not 0 °K nor 273 °K.

As an example of how the other units may be derived from the basic units, velocity is length divided by time. It is usual to write these dimensional equations using square brackets, with the basic quantities being in capital letters and the derived quantities being in lower case letters. Thus for this example we can write $[v] = [L]/[T]$, or

$$[v] = [L\,T^{-1}] \tag{1.1}$$

Example 1.1

Obtain the dimensions of (1) acceleration, (2) force, (3) torque.

Solution

1 Acceleration is the rate of change of velocity, so is velocity/time. Thus

$[a] = [v]/[T]$. From Equation (1.1) we have that $[v] = [L\,T^{-1}]$, so

$$[a] = [L\,T^{-1}]/[T] = [L\,T^{-1}][T^{-1}]$$

or

$$[a] = [L\,T^{-2}] \tag{1.2}$$

2. Force is mass times acceleration. Thus $[f] = [M][a]$. From Equation (1.2) we have that $[a] = [L\,T^{-2}]$, so

$$[f] = [M][L\,T^{-2}]$$

or

$$[f] = [M\,L\,T^{-2}] \tag{1.3}$$

3. Torque is force times the length of the torque arm. Thus $[t] = [f][L]$. From Equation (1.3) we have that $[f] = [M\,L\,T^{-2}]$, so

$$[t] = [M\,L\,T^{-2}][L]$$

or

$$[t] = [M\,L^2\,T^{-2}] \tag{1.4}$$

Example 1.2

Determine the dimensions of (1) energy, (2) power.

Solution

1. Energy is work, which is force multiplied by distance. Thus $[w] = [f][L]$. From Equation (1.3) we have that $[f] = [M\,L\,T^{-2}]$, so

$$[w] = [M\,L\,T^{-2}][L]$$

or

$$[w] = [M\,L^2\,T^{-2}] \tag{1.5}$$

2. Power is energy divided by time. Thus $[p] = [w]/[T]$. From Equation (1.5) we have that $[w] = [M\,L^2\,T^{-2}]$, so

$$[p] = [M\,L^2\,T^{-2}]/[T] = [M\,L^2\,T^{-2}][T^{-1}]$$

or

$$[p] = [M\,L^2\,T^{-3}] \tag{1.6}$$

Example 1.3

Find the dimensions of (1) electric charge, (2) electric potential difference.

Solution

1 Electric charge is electric current multiplied by time. Thus [q] = [A][T], so

$$[q] = [A\,T] \tag{1.7}$$

2 When a charge of 1 coulomb is moved through a potential difference of 1 volt the work done is 1 joule of energy, so that electric potential difference is energy divided by electric charge. Thus [pd] = [w]/[q]. From Equation (1.5) we have that [w] = [M L² T⁻²], and from Equation (1.7) we see that [q] = [A T], so

$$[pd] = [M\,L^2\,T^{-2}]/[A\,T] = [M\,L^2\,T^{-2}][A^{-1}\,T^{-1}]$$

or

$$[pd] = [M\,L^2\,T^{-3}\,A^{-1}] \tag{1.8}$$

Example 1.4

Obtain the dimensions of (1) resistance, (2) inductance, (3) capacitance.

Solution

1 Resistance is electric potential difference divided by electric current. From Equation (1.8) the dimensions of electric potential difference are [M L² T⁻³ A⁻¹]. Thus [r] = [M L² T⁻³ A⁻¹]/[A], so

$$[r] = [M\,L^2\,T^{-3}\,A^{-2}] \tag{1.9}$$

2 The magnitude of the emf induced in a coil of inductance L when the current through it changes at the rate of I ampere in t seconds is given by $e = LI/t$, where e is measured in volts and is a potential difference. Thus the dimensions of L are given by [l] = [pd][T]/[A]. From Equation (1.8), [pd] = [M L² T⁻³ A⁻¹], so

$$[l] = [M\,L^2\,T^{-2}\,A^{-2}] \tag{1.10}$$

3 Capacitance (C) is electric charge (Q) divided by electric potential difference (V). From Equation (1.7), [q] = [A T]. From Equation (1.8), [pd] = [M L² T⁻³ A⁻¹]. Thus [c] = [A T]/[M L² T⁻³ A⁻¹], so

$$[c] = [M^{-1}\,L^{-2}\,T^4\,A^2] \tag{1.11}$$

1.3 DIMENSIONAL ANALYSIS

A necessary condition for the correctness of an equation is that it should be dimensionally balanced. It can be useful to perform a dimensional analysis on equations to check their correctness in this respect. This can be done by checking that the dimensions of each side of an equation are the same.

Example 1.5

The force between two charges q_1 and q_2 separated by a distance d in a vacuum is given by $F = q_1 q_2 / 4\pi\epsilon_0 d^2$, where ϵ_0 is a constant whose dimensions are [M L^{-3} T^4 A^2]. Check the dimensional balance of this equation.

Solution

The left-hand side of the equation is simply the force F and from Example 1.1 (2) we see that its dimensions are [M L T^{-2}].

The dimensions of the right-hand side are $[q][q]/[4][\pi][\epsilon_0][d^2]$. From Example 1.3 (1) we see that the dimensions of electric charge $[q]$ are [A T]. Numbers are dimensionless so that the figure 4 and the constant π have no dimensions. We are told in the question that the dimensions of ϵ_0 are [M^{-1} L^{-3} T^4 A^2]. The distance between the charges, d, has the dimensions of length so that d^2 has the dimensions $[L^2]$. The dimensions of the right-hand side of the equation are therefore

[A T][A T]/[M^{-1} L^{-3} T^4 A^2][L^2] = [A^2 T^2][M L^3 T^{-4} A^{-2} L^{-2}] = [M L T^{-2}]

which is the same as that obtained for the left-hand side of the equation. The equation is therefore dimensionally balanced.

Example 1.6

The energy in joules stored in a capacitor is given by the expression $(C^a V^b)/2$, where C is the capacitance of the capacitor in farads and V is the potential difference in volts maintained across its plates. Use dimensional analysis to determine the values of a and b.

Solution

We have that $W = (C^a V^b)/2$
In dimensional terms $[w] = [c]^a [pd]^b$
From Equation (1.5), $[w]$ = [M L^2 T^{-2}]
From Equation (1.11), $[c]$ = [M^{-1} L^{-2} T^4 A^2]
From Equation (1.8), $[pd]$ = [M L^2 T^{-3} A^{-1}] therefore

[M L^2 T^{-2}] = [M^{-1} L^{-2} T^4 A^2]^a[M L^2 T^{-3} A^{-1}]^b

Equating powers of [M], $1 = -a + b \Rightarrow b = a + 1$
Equating powers of [A], $0 = 2a - b \Rightarrow b = 2a$
By substitution, $2a = a + 1$, so

$a = 1$ and $b = 2a = 2$

The values required are therefore $a = 1$; $b = 2$.

1.4 MULTIPLES AND SUBMULTIPLES OF UNITS

There is an enormous range of magnitudes in the quantities encountered in electrical and electronic engineering. For example, electric potential can be lower than 0.000 001 V or higher than 100 000 V. By the use of multiples and submultiples we can avoid having to write so many zeros. Table 1.2 shows their names and abbreviations.

Table 1.2

Multiple	Abbreviation	Value	Submultiple	Abbreviation	Value
exa	E	10^{18}	milli	m	10^{-3}
peta	P	10^{15}	micro	μ	10^{-6}
tera	T	10^{12}	nano	n	10^{-9}
giga	G	10^{9}	pico	p	10^{-12}
mega	M	10^{6}	femto	f	10^{-15}
kilo	k	10^{3}	atto	a	10^{-18}

These are the preferred multiples and submultiples and you will see that the powers are in steps of 3. However, because of their convenience there are some others in common use. For example, deci (d), which is 10^{-1}, is used in decibel (dB); and centi (c), which is 10^{-2}, is used in centimetre (cm). Capital letters are used for the abbreviations of multiples and lower case letters are used for the abbreviations of submultiples. The exception is kilo for which the abbreviation is the lower case k, not the capital K.

Example 1.7

Express 10 seconds in (1) milliseconds, (2) microseconds.

Solution

1 To convert from units to multiples or submultiples of units it is necessary to *divide* by the multiple or submultiple. To find the number of milliseconds in 1 second we simply divide by the submultiple 10^{-3}. Thus 1 second = $1/10^{-3} = 10^3$ milliseconds. In 10 seconds there are therefore $10 \times 10^3 = 10^4$ ms.

2 To find the number of microseconds in 10 seconds we divide by the submultiple 10^{-6}. Thus in 10 s there are $10/10^{-6} = 10^7$ μs.

Example 1.8

Express 1 metre in (1) kilometres, (2) centimetres.

Solution

1 To find the number of kilometres in 1 metre we divide by the multiple 10^3. Thus in 1 m there are $1/10^3 = 10^{-3}$ km.

2 To find the number of centimetres in 1 metre we divide by the submultiple 10^{-2}. Thus in 1 m there are $1/10^{-2} = 10^2$ cm.

Example 1.9

Express (1) 10 mV in volts, (2) 150 kW in watts.

Solution

1 To convert from multiples or submultiples to units it is necessary to *multiply* by the multiple or submultiple. Thus to convert from millivolts to volts we multiply by the submultiple 10^{-3}:

$$10 \text{ mV} = 10 \times 10^{-3} \text{ V} = 10^{-2} \text{ V}$$

2 Similarly, to convert from kilowatts to watts we multiply by the multiple 10^3:

$$150 \text{ kW} = 150 \times 10^3 \text{ W} = 1.5 \times 10^5 \text{ W}$$

Example 1.10

Express (1) 10 mV in MV, (2) 5 km in mm, (3) 0.1 μF in pF, (4) 50 MW in GW.

Solution

1 To convert from millivolts to volts we multiply by the submultiple 10^{-3}. Thus

$$10 \text{ mV} = (10 \times 10^{-3}) \text{ V} = 10^{-2} \text{ V}$$

Then to convert to megavolts we divide by the multiple 10^6. Thus

$$10^{-2} \text{ V} = (10^{-2}/10^6) \text{ MV} = 10^{-2} \times 10^{-6} \text{ MV} = 10^{-8} \text{ MV}$$

Therefore, in 10 mV there are 10^{-8} MV

2 In this case we multiply by the multiple kilo (10^3) and then divide by the submultiple milli (10^{-3}):

$$5 \text{ km} = 5 \times 10^3 \text{ m} = (5 \times 10^3/10^{-3}) \text{ mm} = 5 \times 10^6 \text{ mm}$$

3 First multiply by the submultiple micro (10^{-6}) and then divide by the submultiple pico (10^{-12}):

$$0.1 \text{ μF} = 0.1 \times 10^{-6} \text{ F} = 10^{-7} \text{ F} = (10^{-7}/10^{-12}) \text{ pF} = 10^5 \text{ pF}$$

4 Here we multiply by the multiple mega (10^6) and then divide by the multiple giga (10^9):

$$50 \text{ MW} = 50 \times 10^6 \text{ W} = 5 \times 10^7 \text{ W} = (5 \times 10^7/10^9) \text{ GW} = 5 \times 10^{-2} \text{ GW}$$

1.5 SELF-ASSESSMENT TEST

1. Which of the following are units:
   torque; second; newton; time; kilogram?

2. Which of the following are derived units:
   metre; coulomb; newton; ampere; volt?

3. Obtain the dimensions of magnetic flux (ϕ) given that the emf induced in a coil of N turns in which the flux is changing at the rate of ϕ webers in t seconds is $N\phi/t$ volts.

4. Obtain the dimensions of magnetic flux density, B (magnetic flux per unit area).

5. Determine the dimensions of potential gradient (change in potential with distance (dV/dx))

6. Express:
   (a) 30 mA in amperes
   (b) 25 A in microamperes
   (c) 10 MW in milliwatts
   (d) 25 nC in coulombs
   (e) 150 pF in nanofarads
   (f) 60 MW in gigawatts
   (g) 150 µJ in millijoules
   (h) 220 Ω in kilohms
   (i) 55 MΩ in milliohms
   (j) 100 N in kilonewtons.

1.6 PROBLEMS

1. Determine the dimensions of magnetic field strength (H) which is measured in amperes per metre.

2. The permeability (μ) of a magnetic medium is the ratio of magnetic flux density (B) to magnetic field strength (H). Determine its dimensions.

3. Obtain the dimensions of electric flux which is measured in coulombs.

4. Electric flux density (D) is electric flux per unit area. Obtain its dimensions.

5. Find the dimensions of electric field strength whose unit is the volt per metre.

6. Check the validity of the statement that the volt per metre is equivalent to the newton per coulomb.

7 The permittivity (ϵ) of the medium of an electric field is the ratio of electric flux density (D) to electric field strength (E). Find its dimensions.

8 The power in watts dissipated in a resistance R ohms through which is flowing a current of I amperes is given by $P = I^a R^b$. Use dimensional analysis to obtain the values of a and b.

9 'The energy (in joules) stored in an inductor having an inductance of L henry through which a current of I amperes is flowing is given by $W = (L^2 I)/2$.' Check whether this statement is true using dimensional analysis.

10 The maximum torque of a three-phase induction motor is given by $T_{max} = k E_2^a X_2^b \omega^c$ where k is a constant, E_2 is the rotor-induced emf measured in volts, X_2 is the rotor reactance measured in ohms and ω is the angular frequency measured in radians per second. Determine the values of a, b and c.

11 The force between two similar magnetic poles of strength p webers separated by a distance d metres in a medium whose permeability is μ is given by $F = k p^a \mu^b d^c$. Obtain the values of a, b and c.

12 Given that the power (in watts) is the product of potential difference (in volts) and current (in amperes), obtain a value for the power in megawatts (MW) dissipated in a resistance when the current through it is 0.35 kA and the potential difference across it is 4.15×10^8 μV.

2 Electric circuit elements

2.1 ELECTRICITY

The atoms which make up all things consist of a number of particles including the electron, the proton and the neutron. The others are more of interest to physicists than to engineers. The electron has a mass of 9.11×10^{-31} kg and carries a negative electric charge; the proton has a mass of 1.6×10^{-27} kg and carries a positive electric charge equal in magnitude to the negative charge of the electron; the neutron has the same mass as the proton but carries no electric charge. Apart from the hydrogen atom, which has one electron and one proton but no neutrons, all atoms contain all three of these subatomic particles. Atoms are normally electrically neutral because they have the same number of electrons as they have protons. If some electrons are removed from the atoms of a body, that body becomes positively charged because it will have lost some negative electricity. Conversely, a body which gains electrons becomes negatively charged (if you comb your hair the comb will gain some electrons and your hair will lose some). Positively charged bodies attract negatively charged bodies and repel other positively charged bodies (which is why the comb can make your hair stand on end!).

The total surplus or deficiency of electrons in a body is called its charge. The symbol for electric charge is Q and its SI unit is the coulomb (C) in honour of Charles Coulomb (1736–1806), a French physicist. The smallest amount of known charge is the charge on an electron which is 1.6×10^{-19} C. It follows that 6.25×10^{18} electrons ($1/1.6 \times 10^{-19}$) are required to make up 1 C of charge. When electric charges are in motion they constitute an electric current which we call electricity.

Electricity is a very convenient form of energy. It is relatively easy to produce in bulk in power stations whether they be coal fired, oil fired or nuclear (using steam turbines to drive the generators) or hydro (using water turbines to drive the generators). A modern coal-fired or nuclear power station typically produces 2000 MW using four 500 MW generators driven by steam turbines at 3000 r/min. The steam required to drive the turbines is raised by burning coal or from the heat produced in a nuclear reactor. Once generated it is transmitted, by means of overhead lines or underground cables, to load centres where it is used. Since generation takes place at about 25 kV, transmission at up to 400 kV, and utilization at around 240 V to 415 V, transformation of voltage levels is required and is conveniently carried out using the transformer. Finally, in use it is extremely flexible and most industrial and domestic premises rely heavily on it for lighting and power.

2.2 ELECTRIC CIRCUITS

Electric circuits or networks are the assemblage of devices and or equipment needed to connect the source of energy to the user or the device which exploits it. Communications systems, computer systems and power systems all consist of more or less complicated electric circuits which themselves are made up of a number of circuit elements. The devices and equipment mentioned above may be represented by 'equivalent circuits' consisting of these circuit elements, and an equivalent circuit must behave to all intents and purposes in the same way as the device or equipment which it represents. In other words, if the device were put into one 'black box' and the equivalent circuit were put into another 'black box', an outside observer of the behaviour of each would be unable to say which black box contained the real device and which contained the equivalent circuit. In practice it is virtually impossible to achieve exact equivalence.

2.3 CIRCUIT ELEMENTS

Circuit elements are said to be either active (if they supply energy) or passive and the elements which make up a circuit are:

- a voltage or current source of energy (active elements);
- resistors, inductors and capacitors (passive elements).

Energy sources

There are two basic variables in electric circuits, namely electric current and electric potential difference (which we will often call voltage for short). A source of energy is required to cause a current to flow and thereby to produce electric voltages in various parts of the circuit. Energy is work and is measured in joules (J) in honour of James Prescott Joule (1818–89), a British scientist. When a force (F newtons) moves a body through a distance (d metres) the work done is ($F \times d$) joules.

Example 2.1

Calculate the work done when a force of 10 N moves a body through a distance of 5 m.

Solution

The work done is force times distance moved = $F \times d = 10 \times 5 = 50$ J.

Voltage source

An ideal voltage source is independent of the current through it. Its electromotive force (emf) or voltage is a function of time only. If a thick copper wire were connected across its ends the current through it would be infinite. The symbol for an ideal voltage source is shown in Fig. 2.1.

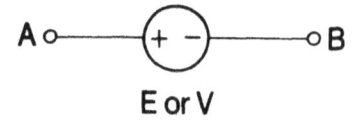

A o———(+ −)———o B

E or V

Figure 2.1

The electric potential difference between two points is defined as being the work required to move a unit positive charge (i.e. 1 C) between them. The unit is called the volt (V) in honour of Alessandro Volta (1745–1827), the Italian inventor of the electric battery. A potential difference of 1 V exists between two points when one joule of work (1 J) is required to move 1 C from the point of lower potential to that of the higher potential.

Example 2.2

Calculate (1) the work done when 300 C of charge is moved between two points having a potential difference of 100 V between them; (2) the potential difference between two points A and B if 500 J of work is required to move 2 mC from A to B.

Solution

1. Work done = charged moved × potential difference through which it is moved
   = QV
   = $300 \times 100 = 30$ kJ

2. Potential difference = work done/charge moved = $500/2 \times 10^{-3} = 250$ kV with point B at the higher potential.

Current source

An ideal current source is independent of the voltage across it and if its two ends are not connected to an external circuit the potential difference across it would be infinite. The symbol for a current generator is shown in Fig. 2.2.

A steady flow of electric charges which does not vary with time is called a

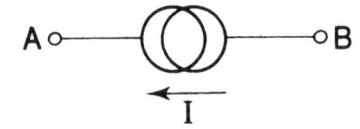

Figure 2.2

direct current. The symbol for current is I and its unit is called the ampere (A) in honour of Andre Ampere (1775–1836), a French mathematician and scientist. When 1 C of charge passes a given plane of reference in one second it represents a current of 1 A, thus

$$I = dQ/dt \quad (I \text{ is the rate of change of charge}) \tag{2.1}$$

It follows that when a current of I amperes flows for T seconds the charge moved is given by

$$Q = \int_0^T I \, dt \tag{2.2}$$

Example 2.3

Calculate (1) the time needed for a current of 10 A to transfer 500 C of charge across a given plane of reference; (2) the current flowing if 200 C of charge passes between two points in a time of 10 s.

Solution

1. From Equation (2.1) we have that $I = dQ/dt$, therefore
   $t = Q/I = 500/10 = 50$ s
2. Again $I = dQ/dt = 200/10 = 20$ A

Resistance

Materials within which charges can move easily are called conductors. Examples of good conductors are copper and aluminium in which electrons can move easily but cannot easily move away from the surface and out of the metal. These materials are said to have a low resistance. Materials within which charges cannot move or can move only with great difficulty are called insulators. These materials are said to have a high resistance, and examples of good insulators are glass and mica.

Ohm's law

Experiment shows that for many conducting materials the current (I) passing through the material from one end to the other is proportional to the potential difference appearing across its ends. Mathematically this is stated as $I \propto V$ or $V \propto I$. We can replace the proportionality sign ($\propto$) by an equality sign if we introduce a constant of proportionality. Thus we write

14 Electric circuit elements

$$V = RI \tag{2.3}$$

where R is the constant of proportionality and is called the resistance of the conducting material. This is known as Ohm's law.

Rearranging Equation (2.3), we obtain the defining equation $R = V/I$ and we note that the unit of resistance is the unit of voltage divided by the unit of current, i.e. the volt per ampere. This is called the ohm, symbol Ω, in honour of Georg Ohm (1787–1854), a German scientist. Materials which obey Ohm's law are known as linear or ohmic materials.

Virtually all devices and equipment have inherent resistance. A circuit element designed specifically to have resistance is called a resistor. There are two circuit symbols commonly used for resistance and either is perfectly acceptable. These are shown in Fig. 2.3 together with the characteristic graph.

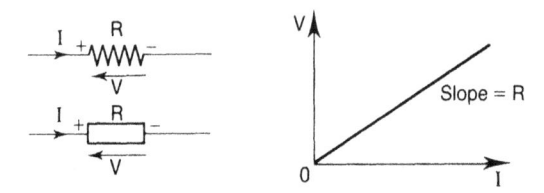

Figure 2.3

The point of entry of the current in a resistor is always positive with respect to the point of exit so far as potential difference is concerned.

Example 2.4

Find the unknown quantities in the diagrams of Fig. 2.4.

Figure 2.4

Solution

(a) Using Ohm's law we have that $R = V/I = 10/5 = 2 \, \Omega$

(b) Again from $V = IR$ we see that V (the voltage across the resistor) $= 5 \times 1 = 5$ V. Since the current enters end A, it is at a higher potential than end B, so $V_B = V_A - 5 = -6 - 5 = -11$ V

(c) Since end B is at a higher potential than end A, the current must enter end B. The potential difference across the resistor is 3 V so that $I = V/R = 3/2 = 1.5$ A, flowing from right to left through the resistor.

Resistivity

The resistance of a conductor is directly proportional to its length (l) and inversely proportional to its cross-sectional area (A). Mathematically then $R \propto l/A$. This may also be written as

$$R = \rho l/A \tag{2.4}$$

where ρ is the constant of proportionality and is called the resistivity of the material of the conductor. Its unit is obtained by rearranging the above equation to make ρ the subject so that $\rho = RA/l$ and we see that the unit of ρ is the unit of R (Ω) multiplied by the unit of A (m^2) divided by the unit of l (m), i.e. $(\Omega\,\text{m}^2)/\text{m} = \Omega\,\text{m}$. The unit of ρ is therefore the ohm-metre. Sometimes it is convenient to use the reciprocal of resistance which is called conductance (G) for which the unit is the siemens (S). Ernst Werner von Siemens (1816–92) was a German inventor. The reciprocal of resistivity is conductivity (σ) for which the unit is the siemens per metre (S m^{-1}). Thus we have that $G = 1/R = A/\rho l$ and since $\sigma = 1/\rho$ we have

$$G = \sigma A/l \tag{2.5}$$

Example 2.5

A copper rod, 20 cm long and 0.75 cm in diameter, has a resistance of 80 $\mu\Omega$. Calculate the resistance of 100 m of wire, 0.2 mm in diameter drawn out from this rod.

Solution

From Equation (2.4), the resistance of the rod is given by $R_R = \rho l_R/A_R$ so that $\rho = R_R A_R/l_R$ where A_R is the cross-sectional area of the rod and l_R is its length. Putting in the values

$$\rho = \{80 \times 10^{-6} \times [\pi(0.0075)^2/4]\}/0.2 = 1.77 \times 10^{-8}\,\Omega\,\text{m}$$

For the wire $R_W = \rho l_W/A_W$ where R_W is the resistance of the wire, A_W is the cross-sectional area of the wire and l_W is its length. Putting in the values,

$$R_W = [1.77 \times 10^{-8} \times 100]/\pi(0.0001)^2 = 56\,\Omega$$

Table 2.1 illustrates the enormous range of values of resistivity (and conductivity) exhibited by various materials. We shall see in the next section that resistance (and resistivity and conductivity) varies with temperature; the values given here are at 20 °C. Remember: the higher the conductivity the better the conductor:

16 *Electric circuit elements*

Table 2.1

Material	Conductivity (S m^{-1})	Resistivity (Ω m)
Silver	6.1×10^7	1.64×10^{-8}
Copper	5.7×10^7	1.75×10^{-8}
Carbon	3×10^4	3.33×10^{-5}
Distilled water	1×10^{-4}	1×10^4
Glass	1×10^{-12}	1×10^{12}
Mica	1×10^{-15}	1×10^{15}
Quartz	1×10^{-17}	1×10^{17}

Resistors in series

If a number of resistors are connected as shown in the diagram of Fig. 2.5 they

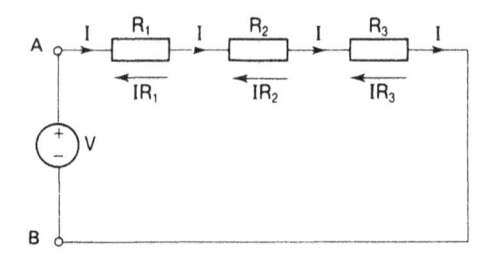

Figure 2.5

are said to be in series. Resistors are in series, therefore, if the same current flows through each of them. In the diagram of Fig. 2.6, for example, only the resistors R_5 and R_6 are in series with each other. Resistor R_1 is in series with the combination of all the others.

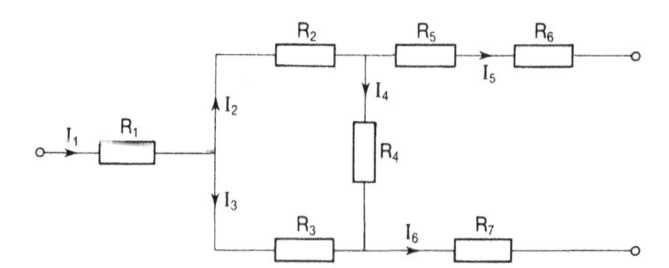

Figure 2.6

By Ohm's law the potential difference across the resistors R_1, R_2 and R_3 in Fig. 2.5 is given by IR_1, IR_2 and IR_3, respectively. The total potential difference between the terminals A and B is therefore $IR_1 + IR_2 + IR_3 = I[R_1 + R_2 + R_3]$. Although this seems obvious, we have, in fact, anticipated Kirchhoff's voltage law which will be stated formally in Chapter 3. A single resistor which would take the same current (I) from the same source (V) would have to have a resistance of $[R_1 + R_2 + R_3]$. The equivalent resistance (R_{eq}) of the three resistors in series is therefore the sum of the three individual resistances. In general, for n resistors is series, $R_{eq} = R_1 + R_2 + \cdots + R_n$. In short this can be written

$$R_{eq} = \sum_{a=1}^{n} R_a \tag{2.6}$$

Example 2.6

Determine (1) the current flowing in the circuit of Fig. 2.7, (2) the voltage across each resistor.

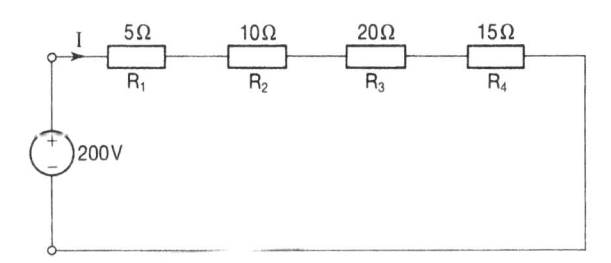

Figure 2.7

Solution

1. Using Ohm's law $I = V/R_{eq}$ and from Equation (2.6), $R_{eq} = R_1 + R_2 + R_3 + R_4$, so $R_{eq} = 5 + 10 + 20 + 15 = 50 \,\Omega$. Therefore $I = 200/50 = 4$ A

2. Again, from Ohm's law $V_{R1} = IR_1 = 4 \times 5 = 20$ V
$V_{R2} = IR_2 = 4 \times 10 = 40$ V
$V_{R3} = IR_3 = 4 \times 20 = 80$ V
$V_{R4} = IR_4 = 4 \times 15 = 60$ V

Note that these add up to 200 V, which is the voltage of the supply.

Voltage division

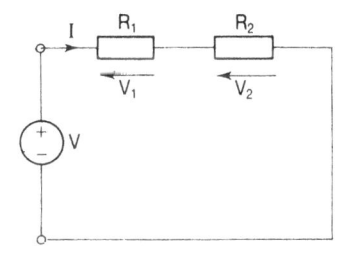

Figure 2.8

For the two resistors shown connected in series in Fig. 2.8, $V = I[R_1 + R_2]$. Also $V_1 = IR_1$ so that

18 Electric circuit elements

$$V_1/V = IR_1/[I(R_1 + R_2)]$$

and

$$V_1 = R_1V/(R_1 + R_2) \tag{2.7}$$

Similarly,

$$V_2 = R_2V/(R_1 + R_2) \tag{2.8}$$

This shows that the ratio of the voltage across a resistor in a series circuit to the total voltage is the ratio of the resistance of that resistor to the total resistance.

Example 2.7

The diagram of Fig. 2.9 shows a variable resistor R_1 in series with a fixed resistor $R_2 = 30 \, \Omega$. Determine (1) the voltage V_2 appearing across R_2 when R_1 is set at $20 \, \Omega$; (2) the value to which R_1 must be set to make the voltage across $R_2(V_2) = 150 \, \text{V}$.

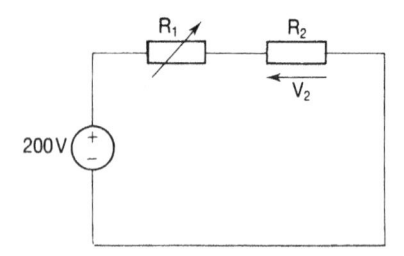

Figure 2.9

Solution

1 From Equation (2.8) we have
  $V_2 = R_2V/(R_1 + R_2) = 30 \times 200/(20 + 30) = 120 \, \text{V}$

2 Rearranging Equation (2.8) to make R_1 the subject, we have
  $R_1 = (R_2V/V_2) - R_2$. Putting in the numbers,
  $R_1 = \{(30 \times 200)/150\} - 30 = 40 - 30 = 10 \, \Omega$

Resistors in parallel

If a number of resistors are connected as shown in Fig. 2.10 they are said to be in parallel. Resistors are in parallel if the same voltage exists across each one.

The total current I is made up of I_1 flowing through R_1, I_2 flowing through R_2 and I_3 flowing through R_3 and by Ohm's law these currents are given by V/R_1, V/R_2 and V/R_3, respectively. It follows that $I = I_1 + I_2 + I_3$ (again this seems obvious but this time we have anticipated Kirchhoff's current law which is formally introduced in Chapter 3). So

2.3 Circuit elements

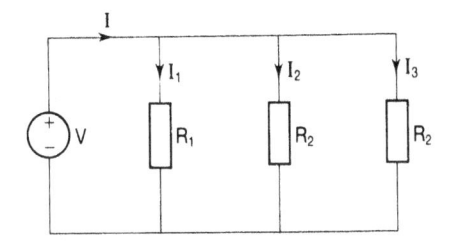

Figure 2.10

$$I = (V/R_1) + (V/R_2) + (V/R_3) = V[(1/R_1) + (1/R_2 + (1/R_3)] \quad (2.9)$$

If a single resistor R_{eq} connected across the voltage source (V) were to take the same current (I) then

$$I = V/R_{eq} \quad (2.10)$$

Comparing Equations (2.9) and (2.10) we see that

$$1/R_{eq} = 1/R_1 + 1/R_2 + 1/R_3$$

In general for n resistors connected in parallel

$$1/R_{eq} = 1/R_1 + 1/R_2 + 1/R_3 + \cdots + 1/R_n \quad (2.11)$$

Since conductance (G) is the reciprocal of resistance ($G = 1/R$) we see that

$$G_{eq} = G_1 + G_2 + \cdots + G_n \quad (2.12)$$

The equivalent conductance of a number of conductances in parallel is thus the sum of the individual conductances.

Example 2.8

Determine the current I flowing in the circuit of Fig. 2.11

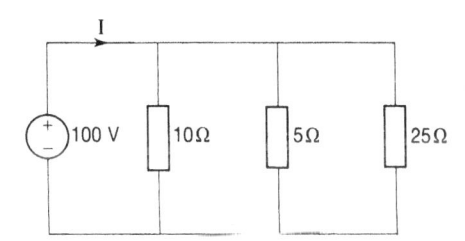

Figure 2.11

Solution

$I = V/R_{eq} = VG_{eq}$ where R_{eq} and G_{eq} are, respectively, the equivalent resistance and conductance of the parallel combination. From Equation (2.12), $G_{eq} = G_1 + G_2 + G_3 = 1/10 + 1/5 + 1/25 = 0.1 + 0.2 + 0.04 = 0.34$ S. Therefore

20 Electric circuit elements

$I = VG_{eq} = 100 \times 0.34 = 34 \text{ A}$

Often we meet just two resistors connected in parallel and it is useful to remember that since $1/R_{eq} = 1/R_1 + 1/R_2 = (R_1 + R_2)/R_1R_2$ then

$$R_{eq} = R_1R_2/(R_1 + R_2) \tag{2.13}$$

i.e. the equivalent resistance of two resistors in parallel is their product divided by their sum.

Current division

In Fig. 2.12 the total current (I) is made up of I_1 flowing through resistor R_1 and

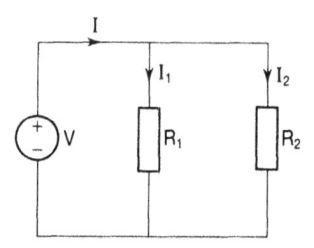

Figure 2.12

I_2 flowing through resistor R_2 and by Ohm's law $I_1 = V/R_1$, $I_2 = V/R_2$ therefore

$$I = I_1 + I_2 = V[(1/R_1) + (1/R_2)]$$
$$= V(R_1 + R_2)/R_1R_2$$
$$I_1/I = (V/R_1)/V(R_1 + R_2)/R_1R_2 = R_2/(R_1 + R_2)$$

So

$$I_1 = R_2I/(R_1 + R_2) \tag{2.14}$$

Similarly

$$I_2 = R_1I/(R_1 + R_2) \tag{2.15}$$

Example 2.9

Determine the current I_2 and the voltage V in the circuit of Fig. 2.13.

Solution

From Equation (2.15), $I_2 = R_1I/(R_1 + R_2) = 10 \times 20/(10 + 40) = 4 \text{ A}$.
From Equation (2.13) $R_{eq} = R_1R_2/(R_1 + R_2) = 400/50 = 8 \text{ }\Omega$

$V = IR_{eq} = 20 \times 8 = 160 \text{ V}$

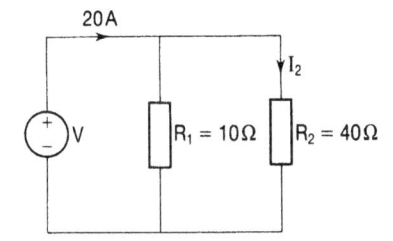

Figure 2.13

Example 2.10

The circuit of Fig. 2.14 is a series–parallel circuit. Calculate (1) the current drawn from the supply (I); (2) the potential difference across the resistor R_4 (V_4); (3) the current through the resistor R_6 (I_6).

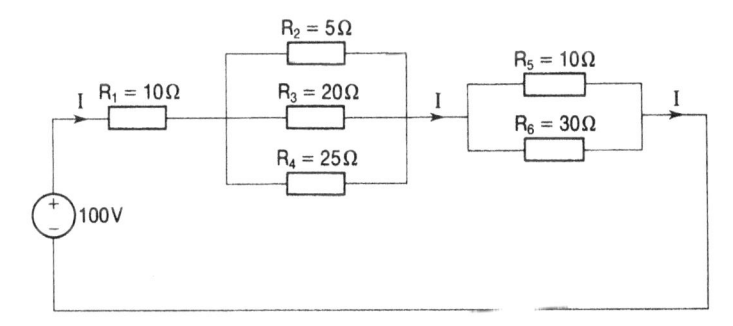

Figure 2.14

Solution

The equivalent resistance of the parallel combination of resistors R_5 and R_6 is given by

$$R_{56} = R_5 R_6/(R_5 + R_6) = 10 \times 30/(10 + 30) = 300/40 = 7.5 \, \Omega$$

For the parallel combination of the resistors R_2, R_3 and R_4 the equivalent resistance is given by

$$1/R_{234} = 1/R_2 + 1/R_3 + 1/R_4 = 0.2 + 0.05 + 0.04 = 0.29 \text{ S}$$

Therefore $R_{234} = 1/0.29 = 3.45 \, \Omega$.

The equivalent resistance of the whole series–parallel circuit is given by $R_{eq} = R_1 + R_{234} + R_{56}$ so $R_{eq} = 10 + 3.45 + 7.5 = 20.95 \, \Omega$

$$I = V/R_{eq} = 100/20.95 = 4.77 \text{ A}$$

$$V_4 = IR_{234} = 4.77 \times 3.45 = 16.46 \text{ V}$$

$$I_6 = R_5 I/(R_5 + R_6) = 10 \times 4.77/40 = 1.19 \text{ A}$$

Internal resistance

It was stated earlier in the chapter that an ideal voltage source is independent of the current flowing through it. Practical voltage sources have internal resistance which means that the voltage at its terminals varies as the current through it changes. The equivalent circuit of a practical voltage source then takes the form shown in Fig. 2.15 where r represents the internal resistance of the source and A and B are its terminals. The terminal voltage is thus V_{AB}.

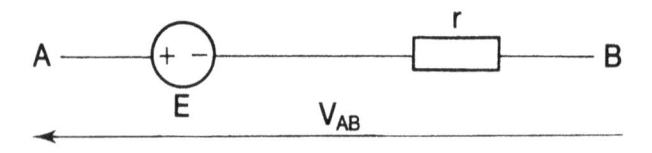

Figure 2.15

Example 2.11

A battery has an internal resistance of 0.5 Ω and a terminal voltage of 15 V when it supplies no current. Determine the terminal voltage when the current through it is 5 A.

Solution

The diagram is as shown in Fig. 2.15. Let the battery terminal voltage when it supplies no current be E (this is called the open circuit voltage). Then, when a current I flows, the terminal voltage $V_{AB} = E - Ir$ where r is the internal resistance. When $I = 5$ A, $V_{AB} = 15 - 5 \times 0.5 = 15 - 2.5 = 12.5$ V.

Effect of temperature

The resistance of metals increases with temperature while for insulators it decreases with temperature. There are some materials for which there is virtually no change in resistance over a wide range of temperatures.

For a given material it is found that

$$R = R_s[1 + \alpha_s(T - T_s)] \tag{2.16}$$

where R is the resistance at a temperature T, R_s is the resistance at temperature T_s, and α_s is the temperature coefficient of resistance corresponding to T_s and is defined as the change in resistance per degree change of temperature divided by the resistance at some temperature T_s. It is measured in $(°C^{-1})$ which is read as 'per degree Celsius'. For a standard temperature $T_s = 0\ °C$, α_s for copper is 0.0043 per °C; for manganin (an alloy of copper, magnesium and nickel) it is 0.000 003 per °C.

If a certain material has a resistance of R_0 at a standard temperature of 0 °C and a resistance temperature coefficient of α_0, then at temperatures T_1 and T_2, respectively, its resistance will be given by $R_1 = R_0[1 + \alpha_0 T_1]$ and $R_2 = R_0[1 + \alpha_0 T_2]$ from which we see that

$$R_1/R_2 = [1 + \alpha_0 T_1]/[1 + \alpha_0 T_2] \tag{2.17}$$

Example 2.12

A copper coil has a resistance of 100 Ω at a temperature of 40 °C. Calculate its temperature at 100 °C. Take α_0 to be 0.0043 per degree C.

Solution

From Equation (2.17) we have that $R_1/R_2 = [1 + \alpha_0 T_1]/[1 + \alpha_0 T_2]$. In this case, $R_1 = 100$ Ω, $T_1 = 40$ °C and $T_2 = 100$ °C. Rearranging Equation (2.17) to make R_2 the subject, we have

$$R_2 = [1 + \alpha_0 T_2]R_1/[1 + \alpha_0 T_1] = [1 + 0.43] \times 100/[1 + 0.172] = 122 \text{ Ω}$$

Colour code for resistors

Some resistors are coded by means of colour bands at one end of the body of the resistor. The first band indicates the first digit of the value of the resistance, the second band gives the second digit and the third band gives the number of zeros. If there is a fourth band this tells us the percentage tolerance on the nominal value. The colour codes are given in Table 2.2.

Table 2.2

First, second and third band		Fourth band (% tolerance)	
black	= 0	gold	= 5
brown	= 1	silver	= 10
red	= 2	none	= 20
orange	= 3		
yellow	= 4		
green	= 5		
blue	= 6		
violet	= 7		
grey	= 8		
white	= 9		

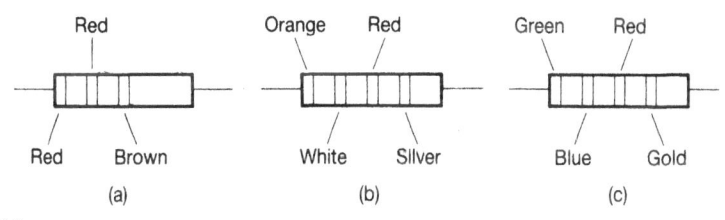

Figure 2.16

(a) Red, Red, Brown
(b) Orange, White, Red, Silver
(c) Green, Blue, Red, Gold

Example 2.13

Write down the nominal value of the resistance of each of the resistors shown in

Fig. 2.16. If they are connected in series, determine the maximum possible resistance of the combination.

Solution

(a) The first band is red so the first digit is 2; the second band is red so the second digit is 2; the third band is brown so there is one zero. There is no fourth band so that the tolerance is 20 per cent. The nominal value of this resistor is therefore 220 Ω and its tolerance is 20 per cent so that its resistance should lie between 220 − 44 = 176 Ω and 220 + 44 = 264 Ω.

(b) The first band is orange so the first digit is 3; the second band is white so the second digit is 9; the third band is red so there are two noughts; the fourth band (silver) means that the tolerance is 10 per cent. The nominal value of this resistor is therefore 3900 Ω (3.9 kΩ) and its value lies between 3510 Ω (−10 per cent) and 4290 Ω (+10 per cent).

(c) The bands on this resistor represent 5 (first digit), 6 (second digit) and red (two zeros) so its nominal value is 5600 Ω (5.6 kΩ). The fourth band (gold) means that its tolerance is ±5 per cent and so its value must be within the range 5320 Ω (5.32 kΩ) to 5880 Ω (5.88 kΩ).

If these resistors were to be connected in series the equivalent resistance of the combination would lie between 9006 Ω (9.006 kΩ) and 10 434 Ω (10.434 kΩ).

Non-linear resistors

A resistor which does not obey Ohm's law, that is one for which the graph of voltage across it to a base of current through it is not a straight line, is said to be non-linear. Most resistors are non-linear to a certain degree because as we have seen the resistance tends to vary with temperature which itself varies with current. So the term non-linear is reserved for those cases where the variation of resistance with current is appreciable. For example, a filament light bulb has a resistance which is very much lower when cold than when at normal operating temperature.

Capacitance

If we take two uncharged conductors of any shape whatever and move Q coulombs of charge from one to the other an electric potential difference will be set up between them (say V volts). It is found that this potential difference is proportional to the charge moved, so we can write $V \propto Q$ or $Q \propto V$. Introducing a constant we have

$$Q = CV \tag{2.18}$$

where C is the constant of proportionality and is called the capacitance of the conductor arrangement. It is a measure of the capacity for storing charge.

An arrangement of conductors having capacitance between them is called a capacitor and the conductors are called plates. The circuit symbol for a capacitor is always as shown in Fig. 2.17 whether the plates themselves are parallel plates, concentric cylinders, concentric spheres or any other configuration of conducting surfaces. The unit of capacitance is the farad (F) named in honour of Michael Faraday (1791–1867), an English scientist.

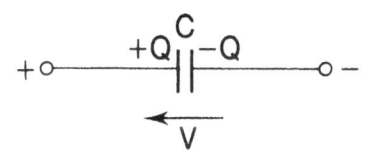

Figure 2.17

It is found that the capacitance of a capacitor depends upon the geometry of its plates and the material in the space between them, which Faraday called the dielectric. For a given arrangement of the plates the capacitance is greater with a dielectric between the plates than it is with a vacuum between them by a factor which is constant for the dielectric. This constant is called the relative permittivity of the dielectric, symbol ϵ_r and is dimensionless. The absolute permittivity (ϵ) of a dielectric is then ϵ_r multiplied by the permittivity of free space (ϵ_0) so that

$$\epsilon = \epsilon_0 \epsilon_r \qquad (2.19)$$

For a vacuum, by definition, $\epsilon_r = 1$.

Permittivity is a very important constant in electromagnetic field theory and relates electric field strength (E) to electric flux density or displacement (D). In fact

$$D = \epsilon E \qquad (2.20)$$

The capacitance of some commonly encountered conductor configurations is given below.

- Parallel plates of cross-sectional area A and separation d:

$$C = A\epsilon/d \quad \text{farad} \qquad (2.21)$$

- Concentric cylinders of radii a (inner cylinder) and b (outer cylinder) of which a coaxial cable is an important example:

$$C = 2\pi\epsilon/\ln(b/a) \quad \text{farad per metre} \qquad (2.22)$$

- Parallel cylinders of radii r and separation d of which overhead transmission lines are an important example:

$$C = \pi\epsilon/\ln(d/r) \quad \text{farad per metre} \tag{2.23}$$

Example 2.14

Two parallel plates each of area 100 cm² are separated by a sheet of mica 0.1 mm thick and having a relative permittivity of 4.

(1) Given that the permittivity of free space (ϵ_0) = 8.854 × 10⁻¹² F/m, calculate the capacitance of the capacitor formed by this arrangement.

(2) Determine the charge on the plates when a potential difference of 400 V is maintained between them.

Solution

From Equation (2.21) the capacitance is given by $C = A\epsilon_0\epsilon_r/d$. In this case $A = 100 \times 10^{-4}$ m²; $d = 0.1 \times 10^{-3}$ m; $\epsilon_r = 4$ and $\epsilon_0 = 8.854 \times 10^{-12}$ F/m. Therefore

$$C = 100 \times 10^{-4} \times 8.854 \times 10^{-12} \times 4/(0.1 \times 10^{-3}) = 3.54 \text{ nF}$$

From Equation (2.18) $Q = CV = 3.54 \times 10^{-9} \times 400 = 1.4$ μC.

Capacitors in series

Capacitors connected as shown in Fig. 2.18 are said to be in series. Applying a voltage V will cause a charge $+Q$ to appear on the left-hand plate of C_1 which will attract electrons amounting to $-Q$ coulombs to the right-hand plate. Similarly, a charge of $-Q$ appears on the right-hand plate of C_2 which will repel electrons from its left-hand plate, leaving it positively charged at $+Q$. Thus the charge throughout this series combination is of the same magnitude (Q). Remember that electric current is charge in motion and that the current at every point in a series circuit is the same. We have seen that $Q = CV$ so that $V_1 = Q/C_1$ and $V_2 = Q/C_2$.

A single capacitor which is equivalent to the series combination would have to have a charge of Q coulombs on its plates and a potential difference of $(V_1 + V_2)$ volts between them. The capacitance of this equivalent capacitor is therefore given by $C_{eq} = Q/V$ and so $V = Q/C_{eq}$. Since $V = V_1 + V_2$ then $Q/C_{eq} = Q/C_1 + Q/C_2$ and

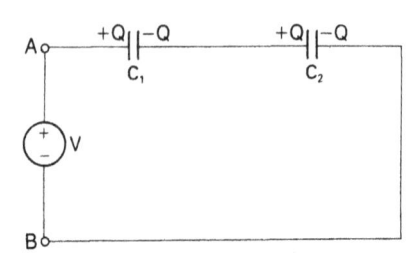

Figure 2.18

$1/C_{eq} = 1/C_1 + 1/C_2$

In general for n capacitors in series we have, for the equivalent capacitance,

$$1/C_{eq} = 1/C_1 + 1/C_2 + \cdots + 1/C_n \tag{2.24}$$

Note that this is of a similar form to the equation for resistors in parallel.

Capacitors in parallel

Capacitors connected as shown in Fig. 2.19 are said to be in parallel. We have

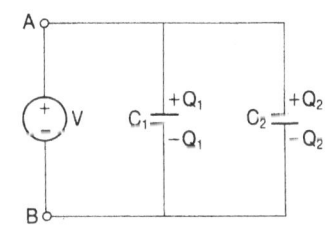

Figure 2.19

that $Q_1 = C_1 V$ and that $Q_2 = C_2 V$. A single capacitor which is equivalent to the parallel combination would have to have a potential difference of V volts between its plates and a total charge of $Q_1 + Q_2$ on them. Thus

$$C_{eq} = (Q_1 + Q_2)/V = (C_1 V + C_2 V)/V = C_1 + C_2$$

In general for n capacitors connected in parallel

$$C_{eq} = C_1 + C_2 + \cdots + C_n \tag{2.25}$$

Note that this is of the same form as the equation for a number of resistors in series.

Example 2.15

Determine the values of capacitance obtainable by connecting three capacitors (of 5 μF, 10 μF and 20 μF) (1) in series, (2) in parallel and (3) in series–parallel.

Solution

Let the capacitors of 5 μF, 10 μF and 20 μF be C_1, C_2, and C_3, respectively.

1. From Equation (2.24) the equivalent capacitance is the reciprocal of $(1/C_1 + 1/C_2 + 1/C_3)$ i.e.
   $1/[(1/5) + (1/10) + (1/20)] = 1/[0.2 + 0.1 + 0.05] = 1/0.35 = 2.86$ μF

2. From Equation (2.25) the equivalent capacitance is
   $C_1 + C_2 + C_3 = 5 + 10 + 20 = 35$ μF

3 (a) When C_1 is connected in series with the parallel combination of C_2 and C_3 the equivalent capacitance is the reciprocal of
$[1/C_1 + 1/(C_2 + C_3)] = 1/[1/5 + 1/30] = 1/[0.2 + 0.033] = 4.29$ μF
(b) Similarly when C_2 is in series with the parallel combination of C_3 and C_1 the equivalent capacitance is
$1/[1/10 + 1/25] = 1/[0.1 + 0.04] = 1/0.14 = 7.14$ μF
(c) Similarly when C_3 is in series with the parallel combination of C_1 and C_2 the equivalent capacitance is
$1/[1/20 + 1/15] = 1/[0.05 + 0.066] = 1/0.116 = 8.62$ μF

Variation of potential difference across a capacitor

From $CV = Q = \int i\, dt$ we have that

$$V = (1/C)\int i\, dt \quad (2.26)$$

It follows that the voltage on a capacitor cannot change instantly but is a function of time.

Inductance

A current-carrying coil of N turns, length l and cross-sectional area A has a magnetic field strength of

$$H = (NI/l) \quad \text{amperes per metre} \quad (2.27)$$

where I is the current in the coil. The current produces a magnetic flux (ϕ) in the coil and a magnetic flux density there of

$$B = (\phi/A) \quad \text{teslas} \quad (2.28)$$

The vectors H and B are very important in electromagnetic field theory.

If the coil is wound on a non-ferromagnetic former or if it is air-cored, then $B \propto H$ and the medium of the magnetic field is said to be linear. In this case

$$B = \mu_0 H \quad (2.29)$$

where μ_0 is a constant called the permeability of free space. Its value is $4\pi \times 10^{-7}$ SI units. If the coil carries current which is changing with time then the flux produced by the current will also be changing with time and an emf is induced in the coil in accordance with Faraday's law. This states that the emf (E) induced in a coil or circuit is proportional to the rate of change of magnetic flux linkages (λ) with that coil ($E \propto d\lambda/dt$). Flux linkages are the product of the flux (ϕ) with the number of turns (N) on the coil, so $E \propto d(N\phi)/dt$. It can be shown that the magnitude of the emf induced in a coil having N turns, a cross-sectional area of A and a length l and which carries a current changing at a rate of dI/dt ampere per second is given by

$$E = [(\mu_0 N^2 A)/l](dI/dt) \quad (2.30)$$

2.3 Circuit elements

The coefficient of dI/dt (i.e. $\mu_0 N^2 A/l$) is called the coefficient of self-inductance of the coil or, more usually, simply the inductance of the coil. A coil having inductance is called an inductor. The symbol for inductance is L and so

$$L = (\mu_0 N^2 A)/l \tag{2.31}$$

Substituting in Equation (2.30) we have

$$E = L(dI/dt) \tag{2.32}$$

From Equation (2.32) we see that the unit of L is the unit of E times the unit of t divided by the unit of I, i.e. the volt-second per ampere (V s A^{-1}). This is called the henry in honour of Joseph Henry (1797–1878), an American mathematician and natural philosopher.

A coil has an inductance of 1 henry when a current changing in it at the rate of 1 ampere per second causes an emf of 1 volt to be induced in it. The circuit symbol for inductance is shown in Fig. 2.20.

Figure 2.20

Non-linear inductance

If the coil is wound on a ferromagnetic former it is found that the flux density B is no longer proportional to the magnetic field strength H (i.e. the flux produced is not proportional to the current producing it). We now write

$$B = \mu H \tag{2.33}$$

where $\mu = \mu_r \mu_0$ and is called the permeability of the medium of the field. It (and μ_r, the relative permeability) varies widely with B. The inductance is now given by

$$L = \mu_0 \mu_r N^2 A/l \tag{2.34}$$

This also varies with B (and H and current) and so is non-linear.

Example 2.16

(1) A wooden ring has a mean diameter of 0.2 m and a cross-sectional area of 3 cm^2. Calculate the inductance of a coil of 350 turns wound on it.

(2) If the wooden ring were replaced by one of a ferromagnetic material having a relative permeability of 1050 at the operating value of magnetic flux density, determine the new value of inductance.

Solution

1. Since the ring, shown in Fig. 2.21, is of wood (a non-ferromagnetic material) the inductance of the coil is given by Equation (2.31) with $N = 350$, $A = 3 \times 10^{-4}$ m^2, $l = \pi \times$ the mean diameter (d) of the coil. Also $\mu_0 = 4\pi \times 10^{-7}$ H/m, therefore

$$L = (\mu_0 N^2 A)/l = (4\pi \times 10^{-7} \times 350^2 \times 3 \times 10^{-4})/0.2\pi = 73.5 \times 10^{-6} \text{ H}$$

2. For the ferromagnetic ring we have, from Equation (2.34), that $L = \mu_0 \mu_r N^2 A/l$. This is just μ_r times the value in part (1). Thus

$$L = 1050 \times 73.5 \times 10^{-6} = 77.18 \times 10^{-3} \text{ H}$$

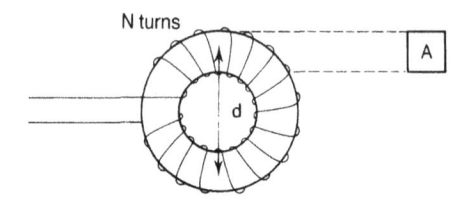

Figure 2.21

Change of current in an inductor

Since $E = L \, dI/dt$ it follows that

$$I = (1/L) \int E \, dt \tag{2.35}$$

This indicates that the current in an inductor is a function of time and therefore cannot change instantaneously. Remember that, in a capacitor, the *voltage* cannot change instantaneously.

Mutual inductance

The diagram of Fig. 2.22 shows two coils placed such that some of the flux produced by a current in either one will link with the other. These coils are said to be mutually coupled magnetically and this is usually indicated in circuit diagrams by a double-headed arrow and the symbol M. Transformer windings are examples of coupled coils.

Let the flux produced by the current i_1 flowing in coil 1 be ϕ_{11} and that part of

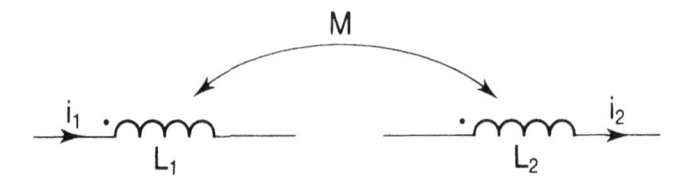

Figure 2.22

it which links coil 2 be ϕ_{21}. Similarly, let the flux produced in coil 2 be ϕ_{22} and that part of it which links coil 2 be ϕ_{12}. The dots are used to indicate the sense of the winding. Thus if current enters the dotted end of coil 1 it will produce a magnetic flux in the same direction as that produced by the current in coil 2 when it enters its dotted end.

If the current in coil 1 is changing with time then the fluxes ϕ_{11} and ϕ_{21} will also change with time. In accordance with Faraday's law, therefore, an emf will be induced in coil 1 because the flux linking it is changing. The magnitude of this emf is given by

$$E_{11} = d(N_1\phi_{11})/dt \tag{2.36}$$

and is called a self-induced emf because it is due to the current changing in the coil itself. Similarly, the changing flux linkages with coil 2 cause an emf to be induced in it and the magnitude of this is given by

$$E_{21} = d(N_2\phi_{21})/dt \tag{2.37}$$

and is called a mutually induced emf because it is caused by the current changing in another coil. We saw (Equation (2.32)) that the self-induced emf is also given by $E_{11} = L_1 di_1/dt$ where L_1 is the self-inductance of coil 1.

Similarly the mutually induced emf in coil 2 may be expressed as

$$E_{21} = M_{12} di_1/dt \tag{2.38}$$

where M_{12} is called the mutual inductance between the coils 1 and 2. If there are only two coils involved there is no need for the double subscript and we can simply write $E_{21} = M di_1/dt$. If the current in coil 2 is changing with time then there will be a self-induced emf E_{22} set up in it and a mutually induced emf E_{12} set up in coil 1 and these are given by

$$E_{22} = L_2 di_2/dt \tag{2.39}$$

$$E_{12} = M di_2/dt \tag{2.40}$$

Coefficient of coupling

If a lot of the flux produced in one coil links with another coil the coils are said to be closely coupled, whereas if only a small amount links, the coils are loosely coupled. It can be shown that for two coils of self-inductance L_1 and L_2 placed such that the mutual inductance between them is M, then

$$M = k\sqrt{(L_1 L_2)} \tag{2.41}$$

where k is called the coefficient of coupling. If $k \rightarrow 1$ the coils are closely coupled whereas if $k \rightarrow 0$ the coils are loosely coupled. If two coils are placed with their magnetic axes at right angles to each other then there is no magnetic coupling between them and k is virtually zero.

Example 2.17

Calculate the mutual inductance between two coils having self-inductances of 2.5 mH and 40 mH if

(1) they are so placed that the coefficient of coupling is 0.8;

(2) one of the coils is wound closely on top of the other;

(3) the coils are places as shown in Fig. 2.23.

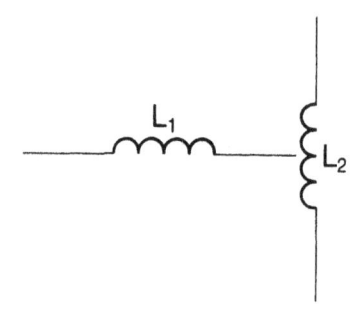

Figure 2.23

Solution

1 From Equation (2.41) we have that

$$M = k\sqrt{(L_1L_2)} = 0.8\sqrt{(2.5 \times 40)} = 8 \text{ mH}$$

2 Since the coils are wound one on top of the other, then virtually all the flux produced will link with both coils and so $k = 1$. Thus

$$M = k\sqrt{(L_1L_2)} = \sqrt{(2.5 \times 40)} = 10 \text{ mH}$$

3 In this case the magnetic axes of the two coils are at right angles so that there is no magnetic coupling and so $k = 0$ and $M = 0$.

Inductance in series

The diagram of Fig. 2.24 shows two coils connected in series electrically and coupled magnetically. The total emf induced in coil 1 is the sum of the self-induced emf due to the current changing in itself and the mutually induced emf due to the current changing in coil 2. Thus

$$E_1 = E_{11} + E_{12} = L_1 di/dt + M di/dt = (L_1 + M) di/dt \tag{2.42}$$

Similarly

$$E_2 = E_{22} + E_{12} = L_2 di/dt + M di/dt = (L_2 + M) di/dt \tag{2.43}$$

The total emf induced in the series combination is therefore given by

$$E_1 + E_2 = (L_1 + L_2 + 2M) di/dt \tag{2.44}$$

2.3 Circuit elements

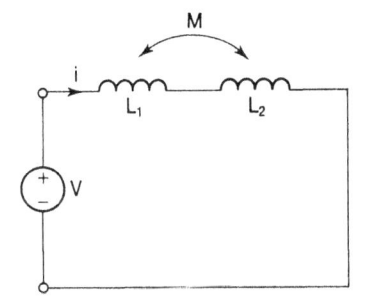

Figure 2.24

This assumes that the coils are wound such that their fluxes are additive (i.e. in the same direction). In this case the coils are said to be connected in series aiding.

If the connections to one of the coils were reversed the flux produced by it would be reversed and the total emf in coil 1 would be $(L_1 - M)di/dt$ while that in coil 2 would be $(L_2 - M)di/dt$ and the total emf in the series combination would be

$$E_1 + E_2 = (L_1 + L_2 - 2M)di/dt \tag{2.45}$$

In this case the coils are said to be connected in series opposing.

A single coil which would take the same current from the same supply as the series aiding combination would need to have an inductance equal to

$$(L_1 + L_2 + 2M) \quad \text{henry} \tag{2.46}$$

and this is called the effective inductance of the circuit. Similarly, the effective inductance of the series opposing combination is

$$(L_1 + L_2 - 2M) \quad \text{henry} \tag{2.47}$$

Example 2.18

Calculate the effective inductance of the two coils arranged as in Example 2.17 (1), (2) and (3) if they are connected in (1) series aiding and (2) series opposing.

Solution

1 Series aiding.
   From expression (2.46) the effective inductance is $(L_1 + L_2 + 2M)$. Now $L_1 = 2.5$ mH and $L_2 = 40$ mH.
   As connected in Example 2.17 part (1) we calculated M to be 8 mH so that the effective inductance is given by $2.5 + 40 + (2 \times 8) = 58.5$ mH.
   As connected in Example 2.17 part (2) we found M to be 10 mH so that the effective inductance becomes $2.5 + 40 + (2 \times 10) = 62.5$ mH.

In Example 2.17 part (3) the M was zero so that the effective inductance is simply $2.5 + 40 = 42.5$ mH.

2 Series opposing.
From expression (2.47), the effective inductance is $(L_1 + L_2 - 2M)$.
For Example 2.17 part (1) this becomes $2.5 + 40 - (2 \times 8) = 26.5$ mH.
For Example 2.17 part (2) it is $2.5 + 40 - (2 \times 10) = 22.5$ mH.
For Example 2.17 part (3) the effective inductance is just
$2.5 + 40 = 42.5$ mH

2.4 LUMPED PARAMETERS

The resistance, capacitance and inductance of transmission lines are not discrete but are distributed over the whole length of the line. The values are then quoted 'per kilometre'. When using equivalent circuit models in such cases the whole of the resistance, capacitance and inductance are often assumed to reside in single elements labelled R, C, and L. These are then called 'lumped parameters'.

Example 2.19

A 50 km three-phase transmission line has the following parameters per phase:

- resistance: 0.5 Ω per kilometre;
- inductance: 3 mH per kilometre;
- capacitance: 16 nF per kilometre.

Draw an 'equivalent circuit' for this line.

Solution

One approximate method of representing this line would be to assume that the whole of the line resistance and inductance is concentrated at the centre of the line and that the whole of the line capacitance is concentrated at one end of the line. This representation is usually quite acceptable for lines of this length because calculations based upon it yield reasonably accurate results.

- The total resistance of the line, $R = 0.5 \times 50 = 2.5$ Ω.
- The total inductance of the line, $L = 3 \times 50 = 150$ mH.
- The total capacitance of the line, $C = 16 \times 50 = 800$ nF $= 0.8$ μF.

If the capacitance is considered to be at the sending end of the line, the equivalent circuit takes the form shown in Fig. 2.25(a); if the capacitance is placed at the receiving end (or load end) of the line, the equivalent circuit is as shown in Fig. 2.25(b).

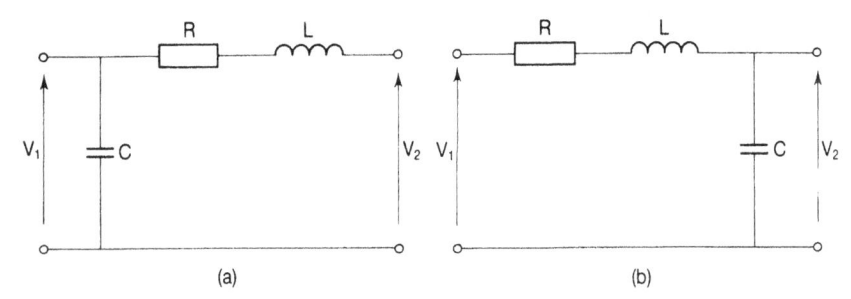

Figure 2.25

2.5 ENERGY STORED IN CIRCUIT ELEMENTS

The circuit elements having inductance and those having capacitance are capable of storing energy. It can be shown that

- the energy stored in a capacitor of C farad across which is maintained a potential difference of V volts is given by

$$W = (CV^2)/2 \quad \text{joules} \tag{2.48}$$

- the energy stored in an inductor of L henry through which a current of I ampere passes is given by

$$W = (LI^2)/2 \quad \text{joules} \tag{2.49}$$

Example 2.20

Determine the energy stored in a capacitor of 0.1 μF when a potential difference of 400 V is maintained across its plates.

Solution

From Equation (2.48) the energy stored is given by

$$W = CV^2/2 = 0.1 \times 10^{-6} \times (400)^2 \times 0.5 = 8 \times 10^{-3} \text{ J}$$

Example 2.21

Calculate the current required to flow through an inductance of 0.5 H in order to store the same amount of energy as that stored by the capacitor in Example 2.20.

Solution

From Equation (2.49) the energy stored is given by $W = LI^2/2$ joules. We know that $W = 8$ mJ (from Example 2.20) and we have to find I. Rearranging the equation to make the current the subject we have $I = \sqrt{(2W/L)}$ amperes. Putting in the numbers,

$$I = \sqrt{(2 \times 8 \times 10^{-3}/0.5 \times 10^{-3})} = \sqrt{32} = 5.66 \text{ A}$$

2.6 POWER DISSIPATED IN CIRCUIT ELEMENTS

Power is the rate of doing work and is measured in watts (W) in honour of James Watt (1736–1819), a British engineer. If we denote power by P and work by W then

$$P = dW/dt \tag{2.50}$$

We may write this as $P = (dW/dQ)(dQ/dt)$. Since work is done when charge is moved through a voltage we have seen above that $W = QV$ so $dW/dQ = V$. Also we have seen (Equation [2.1]) that $dQ/dt = I$. Therefore

$$P = VI \quad \text{watts} \tag{2.51}$$

Using Ohm's law we can also write

$$P = (IR)I = I^2R \tag{2.52}$$

and

$$P = V(V/R) = V^2/R \tag{2.53}$$

Any circuit element having resistance and carrying a current will therefore have an associated power loss given by I^2R watts where R is the resistance in ohms and I is the current in amperes. From Equation (2.50), $P = dW/dt$, from which it follows that energy is power multiplied by time. The energy lost in the element is therefore given by

$$W = I^2Rt \tag{2.54}$$

where t is the time in seconds for which the element is carrying the current I.

Example 2.22

A resistor has a current of 20 A flowing through it and a potential difference of 100 V across it. Calculate (1) the power dissipated in the resistance; (2) the resistance of the resistor; (3) the energy lost in the resistor during each minute of operation.

Solution

1. From Equation (2.51) the power dissipated is
   $P = VI = 100 \times 20 = 2000$ W $= 2$ kW

2. From Equation (2.52), $R = P/I^2 = 2000/20^2 = 5\ \Omega$

3. From Equation (2.54) the energy lost is
   $W = I^2 Rt = 20^2 \times 5 \times 60 = 120\,000\ \text{J} = 120\ \text{kJ}$

Example 2.23

A generator in a power station generates 200 MW of power at 12.7 kV per phase. Calculate the current supplied by the generator.

Solution

From Equation (2.51)

$I = P/V = 200 \times 10^6 / 12.7 \times 10^3 = 15.75\ \text{kA}$

2.7 SELF-ASSESSMENT TEST

1. Define a passive circuit element.

2. State the effect of connecting a copper bar across the terminals of an ideal voltage source.

3. Draw the circuit symbol for an ideal current source.

4. Give two examples of good conductors of electricity and explain why they are good conductors.

5. State Ohm's law.

6. Give the unit of resistivity.

7. What is the reciprocal of conductivity?

8. Calculate the equivalent resistance of three 10 Ω resistors when they are connected (1) in parallel (2) in series and (3) in series–parallel.

9. A voltage source has an open circuit terminal voltage of 15 V and a terminal voltage of 12 V when it supplies a current of 20 A to a load connected across it. Determine the internal resistance of the source.

10. Explain the effect of an increase in temperature upon the resistance of a resistor.

11. A resistor has four colour coded bands as follows: red; red; brown; silver. Between what limits does the resistance of this resistor lie?

12. Explain what is meant by 'a non-linear' resistor.

13. Upon what factors does the capacitance of a capacitor depend?

14 Give an expression for the equivalent capacitance of a number of capacitors connected in series.

15 Can the current change suddenly in a capacitor?

16 Upon what factors does the inductance of an inductor depend?

17 Under what circumstances is the inductance of a coil variable?

18 Can the current through an inductor change suddenly?

19 Two coils of inductance 40 μH and 10 μH are placed such that there is a coefficient of coupling of 0.8 between them. Determine the mutual inductance between them.

20 If the two coils of Question 19 were connected in series-aiding electrically, what would be the effective inductance of the combination?

21 Explain the meaning of lumped parameters.

22 Give an expression for the energy stored in a capacitor in terms of its capacitance and the potential difference across its plates.

23 Give an expression for the energy stored in an inductor in terms of its inductance and the current passing through it.

24 Give an expression for the power dissipated in a resistor in terms of its resistance and the current passing through it.

25 Give the relationship between energy and power.

2.8 PROBLEMS

1 Determine the equivalent resistance of four resistors connected in parallel if their resistances are 1 Ω, 2 Ω, 2.5 Ω and 10 Ω.

2 A 20 Ω resistor is connected in series with a 40 Ω resistor and the combination is connected in series with three 12 Ω resistors which are connected in parallel. Determine the equivalent resistance of the whole arrangement.

3 Calculate the resistance of a 200 m length of copper wire of diameter 1 mm. The resistivity of the copper is 0.0159 μΩ m.

4 Two resistors ($R_1 = 5$ Ω and $R_2 = 20$ Ω) are connected in series across a 100 V supply. Determine the voltage across R_1.

5 If the two resistors in Problem 4 are connected in parallel across the same supply, determine the current through R_2.

6 A battery has an open circuit terminal voltage of 24 V. When it supplies a

current of 2 A the terminal voltage drops to 22 V. Determine the internal resistance of the battery.

7 The winding of a motor has a resistance of 98 Ω at a temperature of 16 °C. After operating for several hours the resistance is measured to be 114 Ω. Determine the steady state operating temperature of the winding. Take the temperature coefficient of resistance to be 0.004 per °C.

8 A 2.2 kΩ resistor has a tolerance of 10 per cent. What are the colour bands on the body of this resistor?

9 A resistor having colour bands orange, orange, brown and silver is connected in parallel with one with bands of yellow, violet, red and gold. Determine the limits of resistance values of the combination.

10 Capacitors of 5 μF, 10 μF and 20 μF are connected in series–parallel in all possible ways. Calculate the values of capacitance obtainable.

11 A wooden ring having a mean diameter of 16 cm and a cross-sectional area of 2 cm^2 is uniformly wound with 500 turns. A second coil of 400 turns is wound over the first such that the coefficient of coupling is unity. Calculate the inductance of each coil and the mutual inductance between them.

12 Calculate the two possible values of effective inductance obtainable by connecting the two coils of Problem 11 in series electrically.

13 Two coils having self-inductances of 100 μH and 50 μH are placed such that the mutual inductance between them is 65 μH. Determine the coefficient of coupling.

14 The energy stored in a coil having an inductance of 30 μH is 1.215 mJ. Determine the current in the coil.

15 A capacitor having a capacitance of 0.1 μF has 200 V maintained across its plates. Determine the energy stored in it.

16 A current of 6 A is passed through a resistor having a resistance of 40 Ω. Determine the power dissipated in the resistor.

3 DC circuit analysis

3.1 INTRODUCTION

Circuit analysis is important in order to be able to design, synthesize and evaluate the performance of electric circuits or networks. The two basic laws for circuit analysis are Kirchhoff's current law (KCL), sometimes referred to as the first law and Kirchhoff's voltage law (KVL), sometimes called the second law. However, a number of techniques have been developed, in the form of network theorems, for simplifying the analysis in the case of more complicated circuits. These theorems, which are introduced in this chapter, are applicable to linear circuits, both a.c. and d.c., but it is convenient to consider d.c. circuits only to begin with because they are a little simpler mathematically and the concepts are that much easier to grasp.

When you have studied this chapter you should be able to calculate the current, voltage and power in any element of most commonly encountered d.c. circuits.

3.2 DEFINITION OF TERMS

It will be useful first of all to define terms, and Fig. 3.1 will be used for this purpose. It shows a five-element circuit of which one (the battery or voltage source) is active and the other four (resistors) are passive.

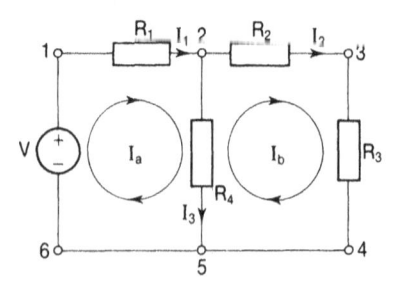

Figure 3.1

- *Node:* a point at which two or more elements have a common connection is called a node. Thus there are six nodes in the circuit, numbered 1–6.

- *Short circuit:* the connection between nodes 4 and 5 is made with a piece of wire having virtually no resistance and is called a short circuit. The connection between nodes 5 and 6 is also a short circuit so that nodes 4, 5, and 6 are identical nodes and the circuit may be redrawn as shown in Fig. 3.2.

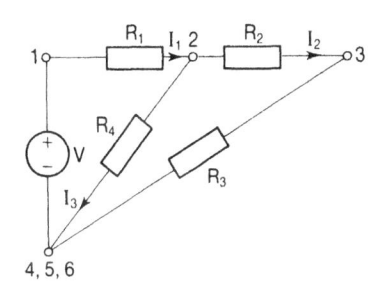

Figure 3.2

In a circuit diagram it would be neater to draw the circuit of Fig. 3.1.

- *Open circuit:* if the resistor R_1 were removed from the circuit then there is said to be an open circuit between nodes 1 and 2. Note that it is incorrect to say that there is then no resistance between nodes 1 and 2 because in fact there is infinite resistance between them.

- *Branch:* a single element or group of elements with two terminals which form the only connections to other single elements or groups of elements is called a branch. In the diagrams of Figs 3.1 and 3.2 there are three branches, one between nodes 6 and 2, one between nodes 2 and 4 and the other between nodes 5 and 2.

- *Branch current:* the current flowing in a branch is called a branch current. Currents I_1, I_2 and I_3 in the diagrams are branch currents.

- *Mesh:* a path through two or more branches which forms a closed path is called a mesh. There are two meshes in the diagram: one passes through resistors R_1, R_4 and the battery V (nodes 1, 2, 5, 6, 1); the other passes through resistors R_2, R_3, and R_4 (nodes 2, 3, 4, 5, 2). A mesh is also called a loop but it cannot have any other loops inside it. The loop containing resistors R_1, R_2, R_3 and the battery V (nodes 1, 2, 3, 4, 5, 6, 1) is therefore not a mesh. In other words, a mesh is a loop but a loop is not necessarily a mesh.

- *Mesh current:* the currents I_a and I_b are called mesh currents. Note that the branch current I_1 is the same as the mesh current I_a but that the branch current I_3 is the mesh current I_a minus the mesh current I_b.

3.3 KIRCHHOFF'S CURRENT LAW

Kirchhoff's current law may be stated as follows. The sum of the currents entering a node is equal to the sum of the currents leaving that node. This means that the algebraic sum of the currents meeting at a node is equal to zero. Applying the law to the node shown in Fig. 3.3, we see that

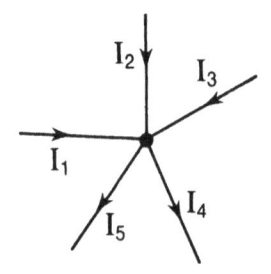

Figure 3.3

$I_1 + I_2 + I_3 = I_4 + I_5$

Rearranging,

$I_1 + I_2 + I_3 - I_4 - I_5 = 0$ (3.1)

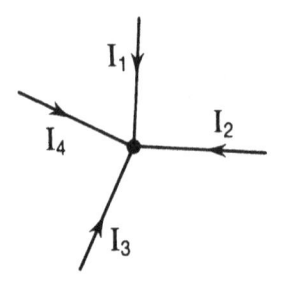

Figure 3.4

For the node shown in Fig. 3.4

$I_1 + I_2 + I_3 + I_4 = 0$ (3.2)

3.4 KIRCHHOFF'S VOLTAGE LAW

Kirchhoff's voltage law may be stated as follows. The sum of the voltage sources around any closed path is equal to the sum of the potential drops around that path. This means that the algebraic sum of all the potential differences around any closed path is equal to zero. The important points to remember are (1) the path must be closed and (2) it is an algebraic sum. Always decide upon a positive direction (say clockwise) for a trip around the path:

potential differences in that direction are then positive and those in the opposite direction are negative.

Applying the law to the circuit of Fig. 3.5 we see that for the closed path containing the nodes 6, 1, 2, 5 and 6, taking the clockwise direction to be positive,

$$I_3R_3 - V + I_1R_1 = 0$$

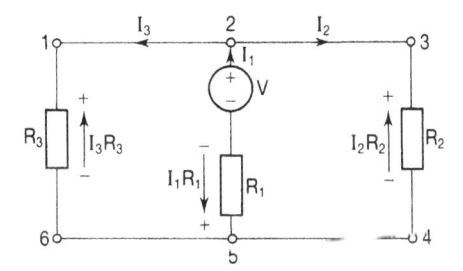

Figure 3.5

rearranging we obtain

$$V = I_1R_1 + I_3R_3 \tag{3.3}$$

For the closed path containing the nodes 5, 2, 3, 4 and 5, taking the clockwise direction to be positive,

$$-I_1R_1 + V - I_2R_2 = 0$$

rearranging we get

$$V = I_1R_1 + I_2R_2 \tag{3.4}$$

Finally, for the closed path containing nodes 6, 1, 2, 3, 4, 5 and 6, taking the clockwise direction to be positive,

$$I_3R_3 - I_2R_2 = 0$$

from which

$$I_2R_2 = I_3R_3 \tag{3.5}$$

Note that Equation (3.5) is not independent because it could have been obtained from Equations (3.3) and (3.4) simply by equating their right hand sides.

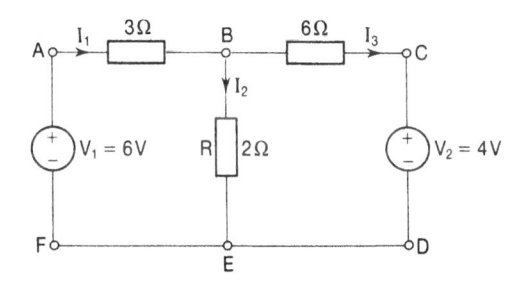

Figure 3.6

Example 3.1

Determine the current flowing in the resistor R in the circuit of Fig. 3.6.

Solution

Let the currents I_1, I_2 and I_3 flow in the three branches as shown. Applying KCL to the node B we see that

$$I_1 = I_2 + I_3 \Rightarrow I_3 = I_1 - I_2$$

We therefore effectively have two unknown currents and we need two independent equations to solve for them. The first of these we obtain by applying KVL to the closed path FABEF which gives, taking the clockwise direction around the path to be positive,

$$V_1 - 3I_1 - 2I_2 = 0$$

Putting $V_1 = 6$ and rearranging, we get

$$3I_1 + 2I_2 = 6 \tag{3.6}$$

The second equation is obtained by applying KVL to the closed path EBCDE to give

$$2I_2 - 6I_3 - V_2 = 0$$

Putting $V_2 = 4$, $I_3 = I_1 - I_2$ and rearranging, we have

$$2I_2 - 6(I_1 - I_2) = 4$$
$$2I_2 - 6I_1 + 6I_2 = 4$$
$$-6I_1 + 8I_2 = 4 \tag{3.7}$$

Since we wish to determine I_2 it is convenient to eliminate I_1. We can do this by multiplying Equation (3.6) by 2 and adding it to Equation (3.7). Thus

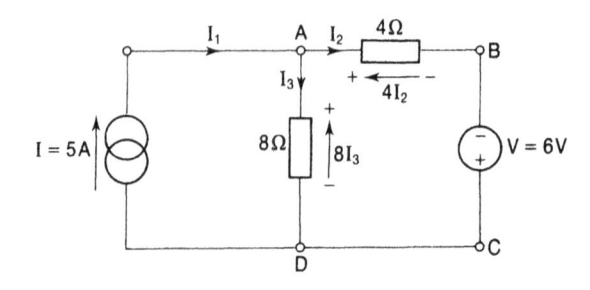

$$6I_1 + 4I_2 = 12 \tag{3.8}$$
$$-6I_1 + 8I_2 = 4 \tag{3.7 bis}$$
$$12I_2 = 16$$
$$I_2 = 1.33 \text{ A}$$

Figure 3.7

Example 3.2

Determine the currents I_3 and I_2 flowing in the circuit of Fig. 3.7.

Solution

Using KCL at node A we see that $I_1 = I_2 + I_3$ so that

$$5 = I_2 + I_3 \text{ and } I_3 = 5 - I_2 \tag{3.9}$$

Applying KVL to the closed path BCDAB and taking the clockwise direction to be positive, we have

$$6 + 8I_3 - 4I_2 = 0 \tag{3.10}$$

Substituting for I_3 from Equation (3.9) above we obtain

$$6 + 8(5 - I_2) - 4I_2 = 0$$
$$6 + 40 - 8I_2 - 4I_2 = 0$$
$$12I_2 = 46$$
$$I_2 = 3.83 \text{ A}$$

It follows that $I_3 = 5 - 3.83 = 1.17$ A.

3.5 THE PRINCIPLE OF SUPERPOSITION

This principle applies to any linear system, and when used in the context of electric circuit theory it may be stated as follows: in any linear network containing more than one source of emf or current, the current in any element of the network may be found by determining the current in that element when each source acts alone and then adding the results algebraically. When removing, in turn, all the sources except one, any voltage source must be replaced by its internal resistance (or by a short circuit if the source is ideal) and any current source must be replaced by an open circuit.

Example 3.3

Calculate the current flowing in the 10 Ω load resistor (R_L) in the circuit of Fig. 3.8.

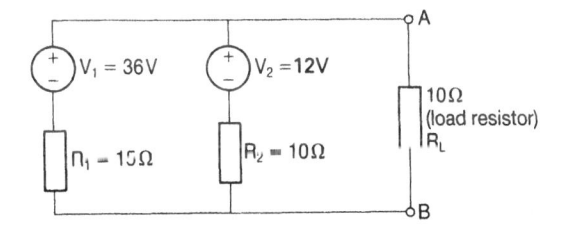

Figure 3.8

Solution

The circuit represents a battery V_1 in parallel with a second battery V_2 supplying a load resistor $R_L = 10\,\Omega$. The terminals A and B are called the load terminals. The resistors R_1 and R_2 represent the internal resistances of the batteries V_1 and V_2, respectively. To apply the principle of superposition, we first remove the battery V_1 to give the circuit of Fig. 3.9.

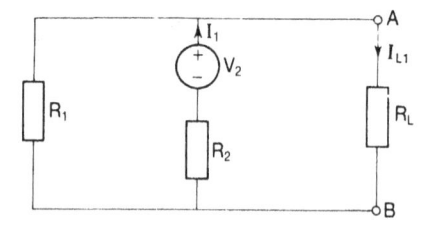

Figure 3.9

Note that the resistor R_2 is in series with the parallel combination of R_1 and R_L so that the current I_1 is given by

$$I_1 = V_2/[R_2 + (R_1 R_L/R_1 + R_L)] = 12/[10 + 150/25)] = 12/16 = 0.75\text{ A}$$

Using the current division technique we see that the current through the load resistor R_L is given by

$$I_{L1} = \{R_1/(R_1 + R_L)\}I_1 = 15 \times 0.75/(15 + 10) = 0.45\text{ A} \quad \text{flowing from A to B}$$

Next we reconnect the battery V_1 and remove the battery V_2, replacing it by its internal resistance (10 Ω). This results in the circuit of Fig. 3.10 in which the

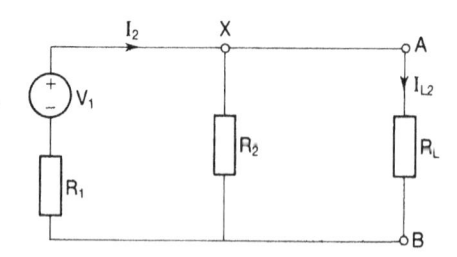

Figure 3.10

resistor R_1 is in series with the parallel combination of R_2 and R_L. The current I_2 is therefore given by

$$I_2 = V_1/[R_1 + (R_2 R_L/(R_2 + R_L)] = 36/[15 + 100/20)] = 36/20 = 1.8\text{ A}$$

At node X this current will divide equally between the resistors R_2 and R_L because they are of equal value. Thus

$$I_{L2} = 0.9\text{ A} \quad \text{flowing from A to B}$$

The current which would flow through the load resistor R_L when both batteries are in the circuit together is therefore, according to the principle of superposition, given by

$I_L = I_{L1} + I_{L2} = 0.45 + 0.9 = 1.35$ A   flowing from A to B

Note that if the battery V_2 were connected in the opposite sense (i.e. with its negative plate connected to the load terminal A and its positive plate connected to B) then I_{L2} would flow from B to A and the load current I_L would be $0.45 - 0.9 = -0.45$ A flowing from A to B (i.e. $+0.45$ A flowing from B to A).

Example 3.4

Determine the current flowing in the 8 Ω resistor (R_1) in the circuit of Fig. 3.11.

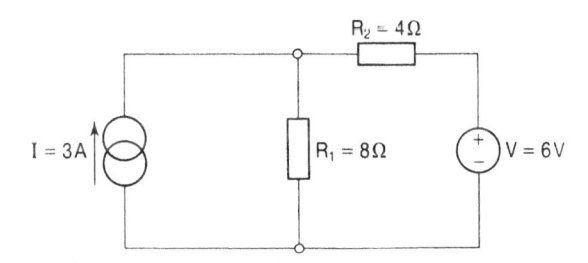

Figure 3.11

Solution

First we replace the current source I by an open circuit, giving the diagram of Fig. 3.12.

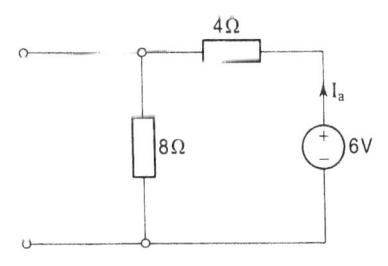

Figure 3.12

From this circuit we see that the current $I_a = V/(R_1 + R_2) = 6/12 = 0.5$ A. Next we reconnect the current source and replace the battery V by a short circuit to give the circuit of Fig. 3.13 shown overleaf.

By current division we obtain

$I_b = [4/(4 + 8)]I = 4 \times 3/12 = 1$ A

When both sources are acting together the current through $R_1 = I_a + I_b = 0.5 + 1 = 1.5$ A

48 DC circuit analysis

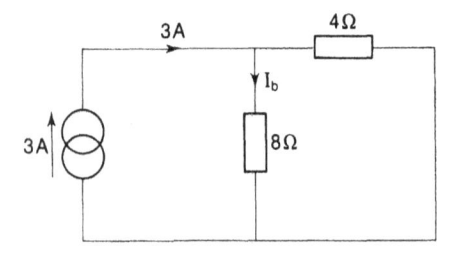

Figure 3.13

3.6 THEVENIN'S THEOREM

Thevenin, a French engineer, developed work by Helmholtz and published this theorem in 1883. It may be stated as follows: any linear network containing an element connected to two terminals A and B may be represented by an equivalent circuit between those terminals consisting of an emf E_0 in series with a resistor R_0.

The emf E_0 is the potential difference between the terminals A and B with the element removed and R_0 is the resistance between the terminals A and B with the element removed and with all sources replaced by their internal resistances. Ideal voltage sources are replaced by a short circuit and ideal current sources are replaced by an open circuit.

In any particular problem, of course, we place the two terminals A and B at either end of the element or part of the circuit in which we wish to determine the current.

Example 3.5

Find the current through the 40 Ω resistor (R_L) in the circuit of Fig. 3.14.

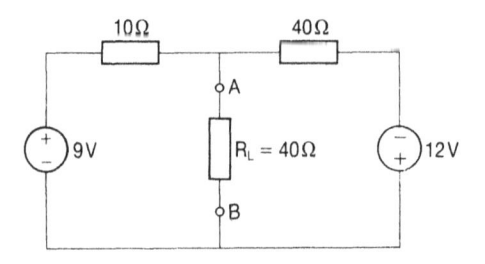

Figure 3.14

Solution

- Step 1: place terminals A and B at either end of the resistor R_L.

- Step 2: represent the circuit by its Thevenin equivalent circuit as shown in Fig. 3.15.

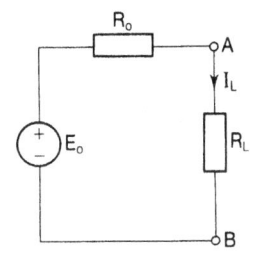

Figure 3.15

- Step 3: to calculate E_0 remove the resistor R_L in Fig. 3.14 to give the circuit of Fig. 3.16 and determine the potential difference between the terminals A and B (say V_{AB}).

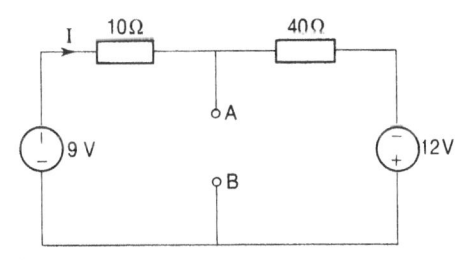

Figure 3.16

To calculate V_{AB} we take a trip from A to B adding the potential drops as we go. We therefore need to calculate the current I. Applying KVI to the circuit and taking the clockwise direction to be positive, we have

$$9 - 10I - 40I + 12 = 0$$
$$50I = 21$$

and

$$I = 21/50 = 0.42 \text{ A}$$

Now, going from A to B via the 9 V battery we have that $-10 \times 0.42 + 9 = 4.8$ V. This means that the potential drop is positive so that terminal A is at a higher potential than terminal B and so the current will flow through R_L from A to B.

To check, we can go from A to B via the 12 V battery in which case we have that $40 \times 0.42 - 12 = 16.8 - 12 = 4.8$ V as before.

- Step 4: to calculate R_0, remove the resistor R_L, replace the batteries by short circuits to give the circuit of Fig. 3.17 on the following page and determine the resistance between the terminals A and B.

    The 10 Ω and the 40 Ω resistors are in parallel so the equivalent resistance between A and B is given by
    $R_0 = (10 \times 40)/(10 + 40) = 8 \text{ Ω}$

50 DC circuit analysis

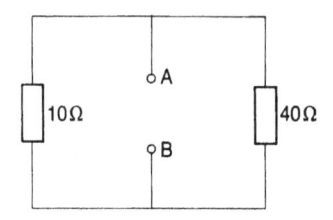

Figure 3.17

- Step 5: put these values for E_0 and R_0 in the Thevenin equivalent circuit of Fig. 3.15. Then

$$I_L = E_0/(R_0 + R_L) = 4.8/(8 + 40) = 0.1 \text{ A}$$

This is therefore the current which will flow through the resistor R_L in the original circuit of Fig. 3.14.

Thevenin's theorem is very useful when we wish to determine the current through or the voltage across an element which is variable.

Example 3.6

The resistor r shown in the diagram of Fig. 3.18 is variable from 0 to 250 Ω. Determine the maximum and minimum values of the current I_L.

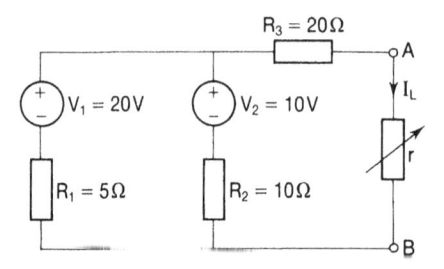

Figure 3.18

Solution

The Thevenin equivalent circuit is shown in Fig. 3.19 and we put terminals A

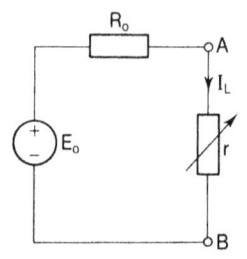

Figure 3.19

3.6 Thevenin's theorem

and B at either end of the resistor, r. To calculate E_0 we first remove the resistor r in the circuit of Fig. 3.18 to give the circuit of Fig. 3.20.

To calculate the potential difference between A and B we need to determine

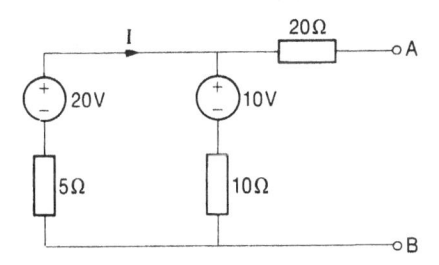

Figure 3.20

the current *I*. Applying KVL to the closed path and taking the clockwise direction to be positive

$$20 - 5I - 10I - 10 = 0$$

$$15I = 10$$

$$I = 0.67 \text{ A}$$

The potential drop between A and B is then given by

$$V_{AB} = IR_2 + V_2 = 0.67 \times 10 + 10 = 16.7 \text{ V}$$

Since this turns out to be positive, then the potential of terminal A is higher than that of terminal B and the current I_L flows from A to B.

In accordance with Thevenin's theorem, $V_{AB} = E_0$.

To calculate R_0 we remove the resistor r in Fig. 3.18 and replace the batteries

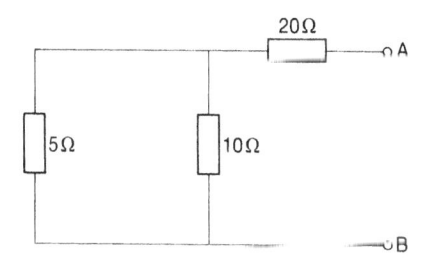

Figure 3.21

by short circuits to give the diagram of Fig. 3.21. R_0 is the resistance between A and B and is given by

$$R_{AB} = R_3 + R_1R_2/(R_1 + R_2)$$
$$= 20 + 50/15$$
$$= 23.33 \text{ }\Omega$$

Now from the Thevenin equivalent circuit of Fig. 3.19, putting in the values for E_0 and R_0,

$$I_L = E_0/(R_0 + r) = 16.7/(23.33 + r) \tag{3.11}$$

I_L is a maximum when r has its minimum value (i.e. 0 Ω):

$$I_{Lmax} = 16.7/23.33 = 0.72 \text{ A}$$

I_L is a minimum when r has its maximum value (i.e. 250 Ω):

$$I_{Lmin} = 16.7/(23.33 + 250) = 0.06 \text{ A}$$

Using Thevenin's theorem the current I_L is now easily obtained for any value of r simply by putting that value into Equation (3.11) above. Using Kirchhoff's laws or the principle of superposition, we would have to rework the whole problem for every value of r.

3.7 NORTON'S THEOREM

In 1926 Norton, an American engineer, introduced an equivalent circuit which is the dual of Thevenin's (duals are discussed in Chapter 10). The theorem may be stated as follows: any linear network containing an element connected to two terminals A and B may be represented by an equivalent circuit between the terminals of a current source I_{SC} in parallel with a resistor R. The current I_{SC} is that which would flow through a short circuit connected between the terminals A and B, and R is the equivalent resistance between them with the element removed, with any voltage source replaced by a short circuit and with any current source replaced by an open circuit.

Example 3.7

Calculate the maximum and minimum values of the potential difference across the resistor r in the circuit of Fig. 3.22 if r is variable between 10 Ω and 100 Ω.

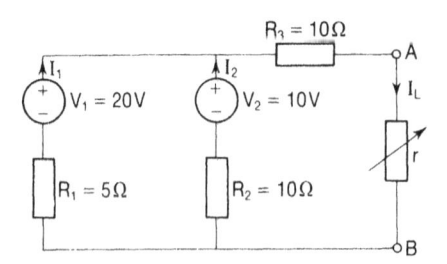

Figure 3.22

Solution

First we put terminals A and B around the resistor r and represent the circuit by its Norton equivalent as shown in Fig. 3.23, from which we see that

$$I_L = [R_{SC}/(R_{SC} + r)]I_{SC} \tag{3.12}$$

3.7 Norton's theorem

and that

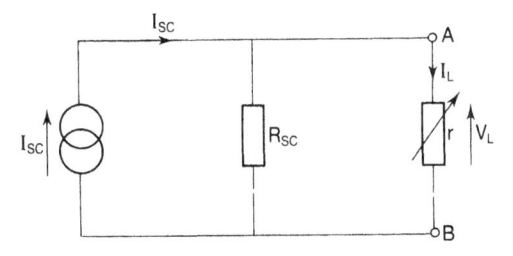

$$V_L = I_L r \tag{3.13}$$

Figure 3.23

To calculate I_{SC} we short circuit the load resistor r as shown in Fig. 3.24 and determine the current through this short circuit. We can make use of the

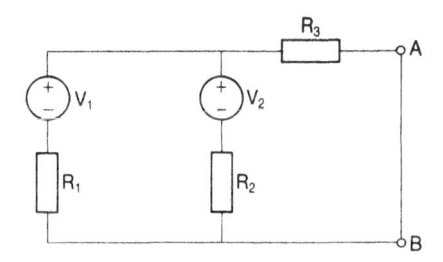

Figure 3.24

principle of superposition to do this. Replace the battery V_2 by a short circuit to give the circuit of Fig. 3.25.

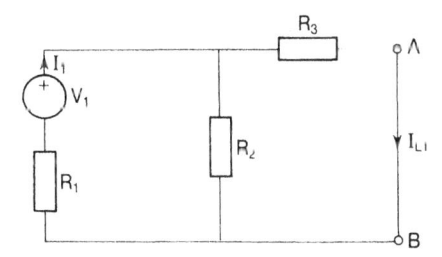

Figure 3.25

From Fig. 3.25 we note that the resistor R_1 is in series with the parallel combination of the resistors R_2 and R_3 so that

$$I_1 = V_1/[R_1 + (R_2 R_3 / R_2 + R_3)] = 20/[5 + (100/20)] = 20/10 = 2 \text{ A}$$

By current division, since $R_2 = R_3$, $I_{L1} = 1$ A.

Now reconnect the battery V_1 and replace the battery V_2 by a short circuit to give the circuit of Fig. 3.26 overleaf. From Fig. 3.26 shown overleaf we see that the resistor R_2 is in series with the parallel combination of the resistors R_1 and R_3 so that

$$I_2 = V_2/[R_2 + (R_1 R_3/(R_1 + R_3)] = 10/[10 + 50/15] = (10/13.33) \text{ A}$$

54 DC circuit analysis

Figure 3.26

By current division

$$I_{L2} = [R_1/(R_1 + R_3)]I_2 = [(5/15) \times (10/13.33)] = 0.25 \text{ A}$$

Now in Equation (3.12),

$$I_{SC} = I_{L1} + I_{L2} = 1 + 0.25 = 1.25 \text{ A}$$

To determine R_{SC} (in Equation (3.12)) we remove r and replace the batteries by short circuits to give the circuit of Fig. 3.27 in which R_{SC} then equals R_{AB}:

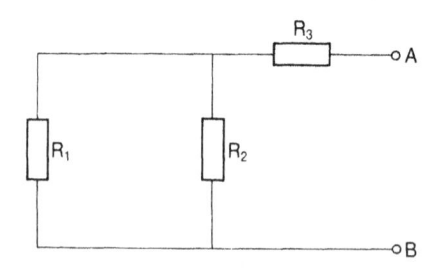

Figure 3.27

$$R_{AB} = R_{SC} = R_3 + R_1R_2/(R_1 + R_2) = 10 + 50/15 = 13.33 \text{ }\Omega$$

The current through the load resistor in Fig. 3.23 is given by

$$I_L = [R_{SC}/(R_{SC} + r)]I_{SC} = [13.33 \times 1.25/(13.33 + r)] \text{ A}$$

According to Norton's theorem this is the current which would flow through r in the circuit of Fig. 3.22.

When $r = 10 \text{ }\Omega$, $I_L = (13.33/23.33)1.25 = 0.71$ A. The corresponding voltage across the load is given by $V_L = 0.71 \times 10 = 7.1$ V.

When $r = 100 \text{ }\Omega$, $I_L = (13.33/113.33)1.25 = 0.15$ A. The corresponding load voltage is then given by $V_L = 0.15 \times 100 = 15$ V.

Note that using this method we need to calculate I_{SC} and R_{SC} only once and, as r varies, simply put its new value into the equation to calculate I_L and V_L.

3.8 THE MAXIMUM POWER TRANSFER THEOREM

Fig. 3.28 shows the Thevenin equivalent circuit of a network. The power in the

3.8 The maximum power transfer theorem

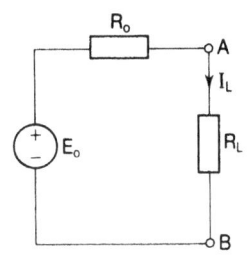

Figure 3.28

load resistor R_L is given by $P_L = I_L^2 R_L$. But $I_L = E_0/(R_0 + R_L)$ so that

$$P_L = E_0^2 R_L/(R_0 + R_L)^2 \tag{3.14}$$

As R_L varies, with E_0 and R_0 being constant, this will be a maximum (or a minimum) when $dP_L/dR_L = 0$. Using the technique for differentiating a quotient we get

$$dP_L/dR_L = \{(R_0 + R_L)^2 E_0^2 - E_0^2 R_L [2(R_0 + R_L)]\}/(R_0 + R_L)^4$$

This will be zero when the numerator is zero, i.e. when

$$(R_0 + R_L)^2 E_0^2 = 2E_0^2 R_L (R_0 + R_L)$$

$$R_0 + R_L = 2R_L$$

$$R_0 = R_L \tag{3.15}$$

This can be confirmed as a maximum, rather than a minimum, by showing that d^2P_L/dR_L^2 is negative.

The power delivered to the load is therefore a maximum when the resistance of the load is equal to the internal resistance of the source or network, and this is called the maximum power transferred theorem. The actual value of the maximum power transferred is obtained by putting $R_L = R_0$ into the equation for P_L. This gives

$$P_{max} = E_0^2 R_L/(R_L + R_L)^2 = E_0^2 R_L/(2R_L)^2$$

$$P_{max} = E_0^2/4R_L. \tag{3.16}$$

Example 3.8

For the circuit of Fig. 3.18 (Example 3.6) determine:

(1) the value of the load resistor, r, which would give the maximum power transfer; and

(2) the maximum power transferred to the load.

Solution

1 Using the Thevenin equivalent circuit of Fig. 3.19, the maximum power

transfer theorem tells us (Equation (3.15)) that the power to the load resistor r will be a maximum when $r = R_0$. In Example 3.6, we found R_0 to be 23.33 Ω. In this case then, for maximum power transfer, $r = 23.33$ Ω.

2 The maximum power transferred will then be given by Equation (3.16) with $R_L = r$

$$E_0^2/4r = (16.7)^2/(4 \times 23.33) = 2.99 \text{ W}$$

3.9 DELTA-STAR TRANSFORMATION

Three resistors connected as shown in Fig. 3.29 are said to be connected in delta (or mesh), while three resistors connected as shown in Fig. 3.30 are said to be star connected. It is often necessary or merely convenient to convert from a delta connection to an equivalent star.

The star circuit will be equivalent to the delta connection if the resistance measured between any two terminals in the star is identical to the resistance measured between the same two terminals in the delta. For this to be the case, the total resistance between terminals 1 and 2 (say R_{12}') in the delta circuit must

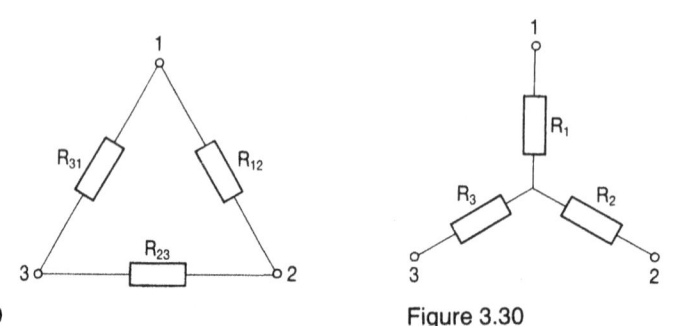

Figure 3.29

Figure 3.30

equal the total resistance $(R_1 + R_2)$ between the same two terminals in the star circuit. Similarly for the resistances between the other two pairs of terminals. We will deal with these in turn.

R_{12}' is the resistance R_{12} in parallel with the series combination of R_{23} and R_{31}, i.e.

$$R_{12}' = R_{12}[R_{23} + R_{31}]/(R_{12} + R_{23} + R_{31})$$

For equivalence, then,

$$R_1 + R_2 = R_{12}[R_{23} + R_{31}]/(R_{12} + R_{23} + R_{31}) \tag{3.17}$$

Similarly, equating the equivalent resistance between terminals 2 and 3 in Fig. 3.30 with that between terminals 2 and 3 in Fig. 3.29:

$$R_2 + R_3 = R_{23}[R_{31} + R_{12}]/(R_{12} + R_{23} + R_{31}) \tag{3.18}$$

3.9 Delta-star transformation

and for the resistance between terminals 3 and 1:

$$R_3 + R_1 = R_{31}[R_{12} + R_{23}]/(R_{12} + R_{23} + R_{31}) \tag{3.19}$$

Subtract Equation (3.18) from Equation (3.17) to give, after expanding the brackets,

$$R_1 - R_3 = (R_{12}R_{23} + R_{12}R_{31} - R_{23}R_{12} - R_{23}R_{31})/(R_{12} + R_{23} + R_{31})$$

$$R_1 - R_3 = (R_{12}R_{31} - R_{23}R_{31})/(R_{12} + R_{23} + R_{31}) \tag{3.20}$$

Adding Equations (3.19) and (3.20), we obtain

$$2R_1 = (R_{31}R_{12} + R_{31}R_{23} + R_{12}R_{31} - R_{23}R_{31})/(R_{12} + R_{23} + R_{31})$$
$$= 2R_{31}R_{12}/(R_{12} + R_{23} + R_{31})$$

So

$$R_1 = R_{31}R_{12}/(R_{12} + R_{23} + R_{31}) \tag{3.21}$$

If we now subtract Equation (3.19) from Equation (3.18) and add the result to Equation (3.17) we obtain

$$R_2 = R_{12}R_{23}/(R_{12} + R_{23} + R_{31}) \tag{3.22}$$

Finally, by subtracting Equation (3.17) from Equation (3.19) and adding the result to Equation (3.18):

$$R_3 = R_{23}R_{31}/(R_{12} + R_{23} + R_{31}) \tag{3.23}$$

An easy way to remember the rules for changing from delta to star is to draw the star set inside the delta set as shown in Fig. 3.31.

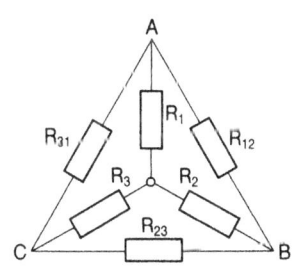

Figure 3.31

Any star equivalent resistor is then given as 'the product of the two delta resistors on either side divided by the sum of all three delta resistors'.

Example 3.9

Determine the input resistance (i.e. the resistance between terminals A and B) in the circuit of Fig. 3.32.

58 *DC circuit analysis*

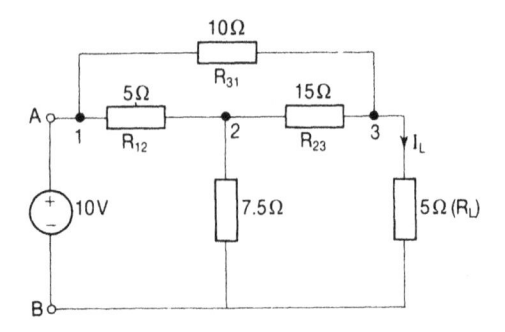

Figure 3.32

Solution

This is called a bridged-T circuit. Note that the resistors R_{12}, R_{23} and R_{31} are connected in delta. This is shown more clearly in Fig. 3.33 in which the star equivalent resistors are shown as R_a, R_b and R_c. The node labelled S is called the star point.

Using the delta-star transformation, we have that

R_a = (product of the adjacent delta resistors/sum of the delta connected resistors)
  $= 10 \times 5/30 = 1.67 \, \Omega$

Similarly

$R_b = 10 \times 15/30 = 5 \, \Omega$

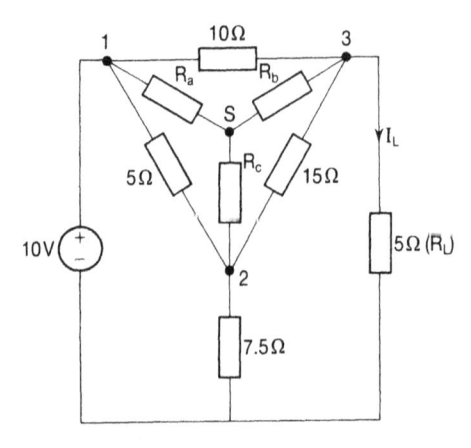

Figure 3.33

And

$R_c = 5 \times 15/30 = 2.5 \, \Omega$

The circuit now simplifies to that of Fig. 3.35 via Fig. 3.34.

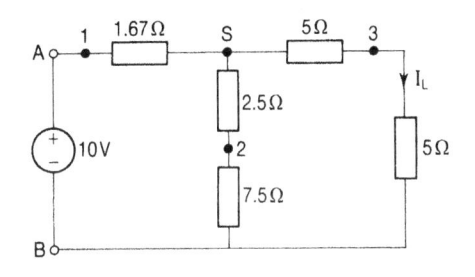

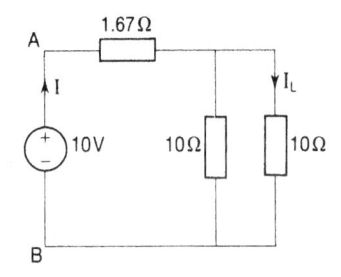

Figure 3.34

Figure 3.35

The equivalent resistance between A and B is then given by

$$R_{AB} = R_a + 10 \times 10/(10 + 10) = 1.67 + 5 = 6.67 \, \Omega$$

The current I_L in Fig. 3.32 is the same as that in Fig. 3.35 and is obtained by current division. Since the two parallel connected resistors are equal in value, the total current I will divide equally between them. Now

$$I = V/R_{AB} = (10/6.67)A = 1.5 \text{ A}$$

so that $I_L = 0.75$ A.

3.10 STAR-DELTA TRANSFORMATION

We can use the same diagrams of Figs 3.29 and 3.30 and obtain the reverse transformation by considering Equations (3.21), (3.22) and (3.23) above. Multiplying Equation (3.21) by Equation (3.22), Equation (3.22) by Equation (3.23) and Equation (3.23) by Equation (3.21) in turn, we obtain

$$R_1 R_2 = R_{12}^2 R_{23} R_{31}/[R_{12} + R_{23} + R_{31}]^2 \tag{3.24}$$

$$R_2 R_3 = R_{23}^2 R_{31} R_{12}/[R_{12} + R_{23} + R_{31}]^2 \tag{3.25}$$

$$R_3 R_1 = R_{31}^2 R_{12} R_{23}/[R_{12} + R_{23} + R_{31}]^2 \tag{3.26}$$

Adding Equations (3.24), (3.25) and (3.26) gives

$$R_1 R_2 + R_2 R_3 + R_3 R_1 = R_{12} R_{23} R_{31} [R_{12} + R_{23} + R_{31}]/[R_{12} + R_{23} + R_{31}]^2$$
$$= R_{12} R_{23} R_{31}/(R_{12} + R_{23} + R_{31})$$

Now $R_{23} R_{31}/(R_{12} + R_{23} + R_{31}) = R_3$ from Equation (3.23) so

$$R_1 R_2 + R_2 R_3 + R_3 R_1 = R_{12} R_3$$

Dividing both sides by R_3 we see that

$$R_{12} = R_1 + R_2 + R_1 R_2/R_3 \tag{3.27}$$

Similarly, by noting that

$$R_{31} R_{12}/(R_{12} + R_{23} + R_{31}) = R_1$$

60 DC circuit analysis

and that

$$R_{12}R_{23}/(R_{12} + R_{23} + R_{31}) = R_2$$

we obtain

$$R_{23} = R_2 + R_3 + R_2R_3/R_1 \qquad (3.28)$$

and

$$R_{31} = R_3 + R_1 + R_3R_1/R_2 \qquad (3.29)$$

Example 3.10

Transform the star connected resistances R_1, R_2 and R_3 shown in the bridged-T network of Fig. 3.36 to a delta. Hence determine the current flowing through the 40 Ω resistor (R).

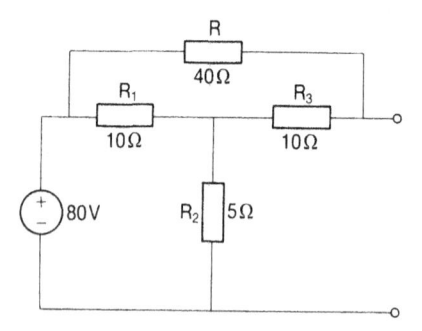

Figure 3.36

Solution

The diagram is redrawn in Fig. 3.37 to show the equivalent delta resistors R_{12},

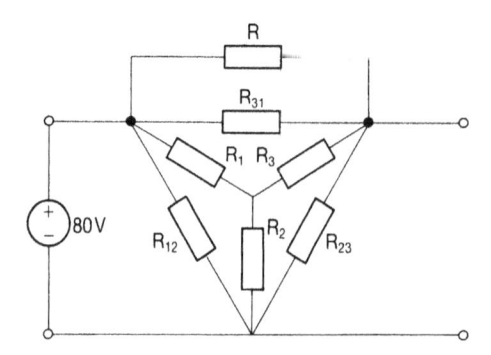

Figure 3.37

R_{23} and R_{31}. Using the star-delta transform Equations (3.27), (3.28) and (3.29), we have

$$R_{12} = R_1 + R_2 + R_1R_2/R_3 = 10 + 5 + 50/10 = 20 \text{ Ω}$$

$R_{23} = R_2 + R_3 + R_2R_3/R_1 = 5 + 10 + 50/10 = 20 \, \Omega$

$R_{31} = R_3 + R_1 + R_3R_1/R_2 = 10 + 10 + 100/5 = 40 \, \Omega$

This leads to the circuit of Fig. 3.38 and then to Fig. 3.39 by replacing the two parallel connected 40 Ω resistors by their 20 Ω equivalent resistor.

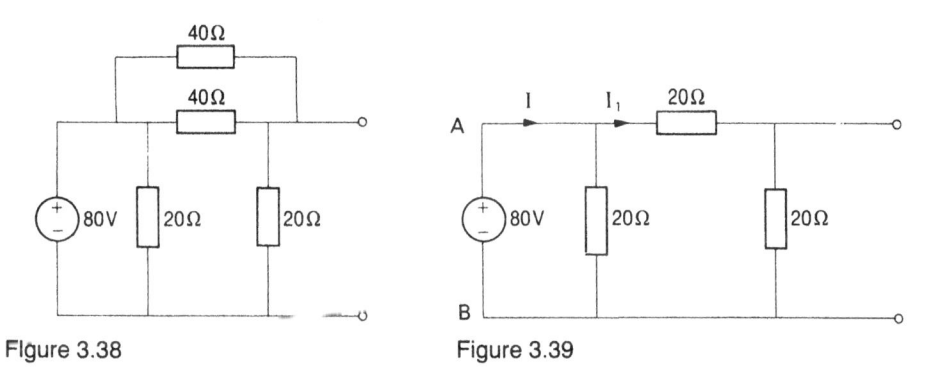

Figure 3.38

Figure 3.39

The equivalent resistance between the terminals A and B in Fig. 3.39 is given by

$R_{AB} = (20 \times 40)/60 = (80/6) \, \Omega$. It follows that

$I = V/R_{AB} = 80 \times 6/80 = 6 \text{A}$.

By current division

$I_1 = (20/60)I = (20/60) \times 6 = 2 \text{ A}$

The current through each 40 Ω resistor in Fig. 3.38 is therefore 1A and since the top one of these is the resistor R in Fig. 3.36, the required answer is 1A.

3.11 SELF-ASSESSMENT TEST

1. State Kirchhoff's first (current) law (KCL).
2. Define a node.
3. Define a mesh and state the difference between a mesh and a loop.
4. State Kirchhoff's second (voltage) law (KVL).
5. Explain the usefulness of the principle of superposition in electrical circuit analysis.
6. State Thevenin's theorem.
7. State Norton's theorem.
8. What is the condition for the maximum power transfer from a source to a load?

62 DC circuit analysis

9 Three resistors each of 12 Ω resistance are connected in delta. Determine the resistance of each of the equivalent star connected resistors.

10 Three resistors each of 10 Ω resistance are connected in star. Calculate the resistance of each of the equivalent delta connected resistors.

11 A voltage source has an internal resistance of 100 Ω and an open circuit terminal voltage of V_S. Determine the resistance of the load resistor if:
(a) the power transferred to the load is to be a maximum;
(b) the required load voltage is $0.9\, V_S$.

12 A voltage source V_S having a resistance R_S supplies a load R_L via a transmission line of resistance R_T. If the power transferred to the load is to be maximized specify:
(a) R_L if $R_S = 100\,\Omega$ and $R_T = 10\,\Omega$
(b) R_T if $R_L = 10\,\Omega$ and $R_S = 5\,\Omega$
(c) R_S if $R_L = 40\,\Omega$ and $R_T = 10\,\Omega$

3.12 PROBLEMS

1 Use (a) Kirchhoff's laws and (b) Thevenin's theorem to determine the current through the 10 Ω resistor in the circuit of Fig. 3.40.

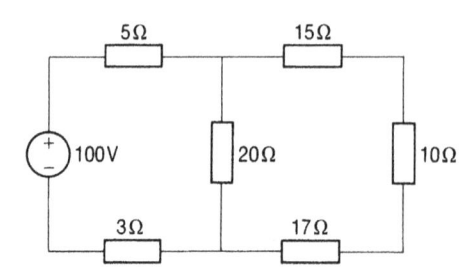

Figuro 3.40

2 Calculate the current in the 1 Ω resistor in the circuit of Fig. 3.41 using (a) the principle of superposition and (b) Thevenin's theorem.

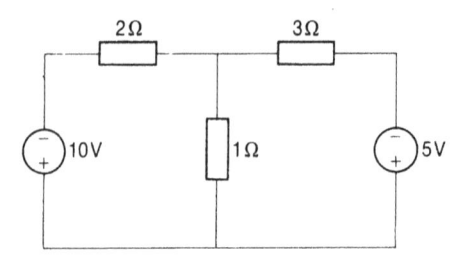

Figure 3.41

3 Determine the Thevenin equivalent circuit between the terminals A and B in the circuit of Fig. 3.42.

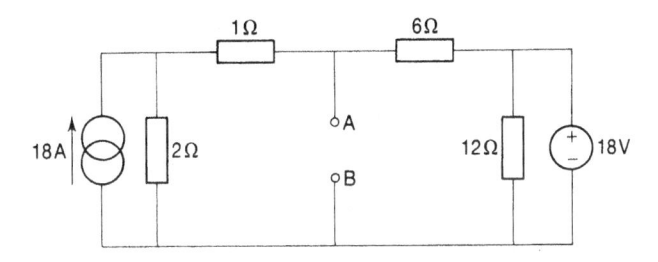

Figure 3.42

4 In the circuit of Fig. 3.43 the resistor R is variable from 0 to 40 Ω. Determine the limits of the current through it.

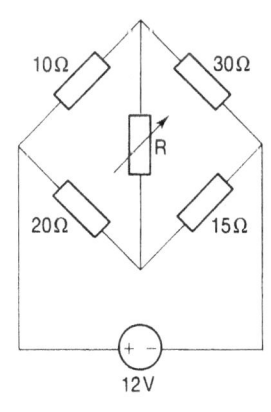

Figure 3.43

5 The resistor R in the circuit of Fig. 3.44 is variable from 2 Ω to 200 Ω. Determine (a) the maximum current through it, (b) the minimum current through it and (c) the value of its resistance when the current through it is 100 mA.

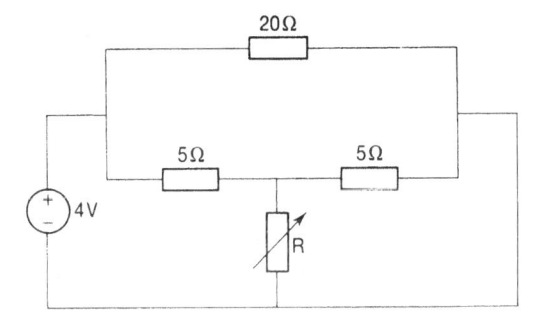

Figure 3.44

6 Use Norton's theorem to calculate the current in each of the branches of the circuit of Fig. 3.45 shown overleaf.

64 *DC circuit analysis*

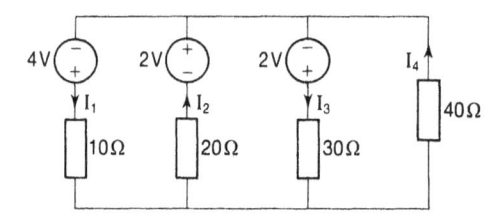

Figure 3.45

7 Three 20 Ω resistors are connected in star and a voltage of 480 V is applied across two of them. If three other resistors (identical to each other) are connected in delta and the same supply is connected across one of them, determine (a) the value of the delta connected resistors and (b) the current in each of the delta connected resistors. The supply current is the same as before.

8 Three resistors having resistances of 100 Ω, 200 Ω and 1 Ω are connected in delta. Determine the resistances of the equivalent star connected resistors.

9 In the circuit of Fig. 3.42, a load resistor (R_L) is connected between terminals A and B. Determine (a) the value of R_L in order that the power transferred to it from the source shall be a maximum and (b) the maximum power transferred.

10 A network has a Thevenin equivalent circuit consisting of a source E_{OC} in series with a resistance R_{OC} so far as its output terminals A and B are concerned. When a load resistor of resistance 25 Ω is connected across the output terminals, the load current is 48.3 mA and when the load resistance is doubled the load current falls to 25.9 mA.
(a) Determine the values of E_{OC} and R_{OC}. (b) Calculate the output voltage when a load of 40 Ω resistance is connected between A and B.

11 A circuit has three nodes A, B and C. Between A and C is connected a voltage source of 20 V (having an internal resistance of 5 Ω) in parallel with a resistor of 15 Ω resistance. Between B and C is connected a voltage source of 10 V (having an internal resistance of 2 Ω) in parallel with a resistor of 8 Ω resistance. The negative terminals of both sources are connected to the node C. Between A and B is connected a resistor, R of 10 Ω resistance. Determine (a) the current through the resistor R, (b) the necessary resistance required in series or in parallel with R for the power to be transferred to terminals AB to be a maximum and (c) the value of the maximum power transferred.

12 A bridge circuit has nodes A, B, C and D. The circuit is arranged as follows:
between A and B: a resistance of 10 Ω

between B and C: a resistance of 80 Ω
between C and D: a resistance of 750 Ω
between D and A: a resistance of 100 Ω
between B and D: a resistance of 200 Ω
between A and C: a voltage source of 20 V having an internal resistance of 4 Ω. The positive terminal of the voltage source is connected to the node A.
Calculate the current through the 750 Ω resistor.
[Hint: convert the delta connection ABD into the equivalent star.]

4 Single-phase a.c. circuits

4.1 ALTERNATING QUANTITIES

A quantity which is continually changing its sign from positive to negative and back again is called an alternating quantity, usually referred to simply as an a.c. quantity. Examples of alternating quantities are shown in Fig. 4.1(a) and (b). The quantities shown in Fig. 4.1(c) and (d) are not alternating but are varying direct quantities.

A graph of the quantity to a base of time is called the waveform of the quantity and when the waveform has completed one complete series of changes and is about to begin to repeat itself it is said to have completed one cycle. The time for one complete cycle is termed the periodic time (T) or simply the period. The number of complete cycles completed in one second is called the frequency (f) and is measured in cycles per second which is called the hertz (Hz) in honour of Heinrich Hertz (1857–94), a German Scientist.

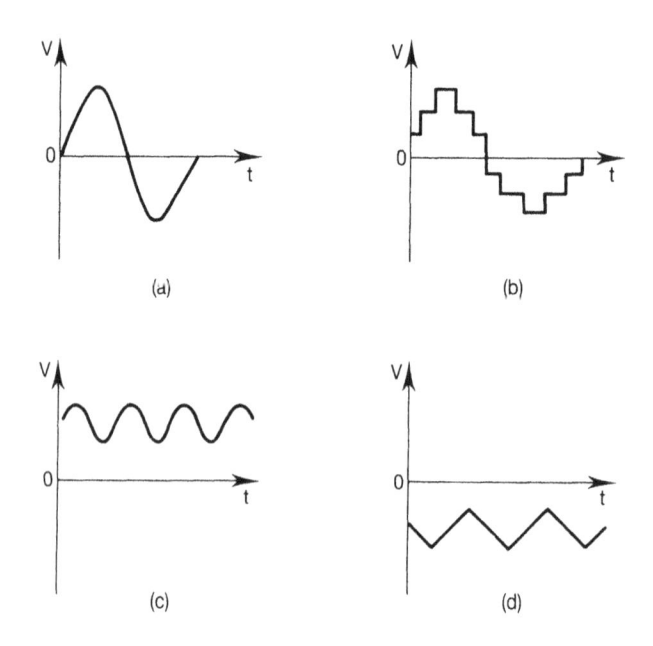

Figure 4.1

It follows that

$$f = 1/T \tag{4.1}$$

Example 4.1

Determine (1) the periodic time of an a.c. quantity of frequency 50 Hz, (2) the frequency of an a.c. waveform for which the period is 2.5 ms.

Solution

1 From Equation (4.1), $f = 1/T$ so that $T = 1/f = 1/50 = 20$ ms.

2 Again from Equation (4.1), $f = 1/T = 1/(2.5 \times 10^{-3}) = 400$ Hz.

There is an enormous range of frequencies and they are banded as shown in Table 4.1.

Table 4.1

Frequency range	Description
– 20 Hz	Low
20 Hz – 15 kHz	Audio
15 kHz – 30 kHz	Very low radio
30 kHz – 300 kHz	Low radio
300 kHz – 3 MHz	Medium radio
3 MHz – 30 MHz	High radio
30 MHz – 300 MHz	Very high (VHF)
300 MHz – 3 GHz	Ultra high (UHF)
3 GHz – 30 GHz	Super high

Instantaneous values

In general an alternating quantity changes its magnitude from instant to instant over the cycle time and these values are called the instantaneous values of the quantity. They are represented by lower case letters, for example i (for current), v (for voltage).

Peak values

The highest value reached by a quantity in the cycle is called the maximum (or peak or crest) value. This value is usually denoted by a capital letter with a circumflex accent or with a subscript max or m so that a peak voltage might be written $\hat{V}$ or V_{max} or V_m.

Sinusoidal a.c. quantities

The beauty of a.c. is that the voltage and current levels can be easily changed by

68 *Single-phase a.c. circuits*

means of a machine called a transformer which, having no moving parts, is extremely efficient. Now the emf induced in transformer windings is proportional to $d\Phi/dt$, the rate of change of magnetic flux linking them (i.e. differentiation is involved). The only a.c. waveform which when differentiated (or integrated) gives the same waveshape is the sine wave. Others become progressively more distorted with each subsequent differentiation, leading to harmonics and reduced efficiency and performance. For this reason the sinusoidal waveform in the most commonly encountered waveform in electrical engineering.

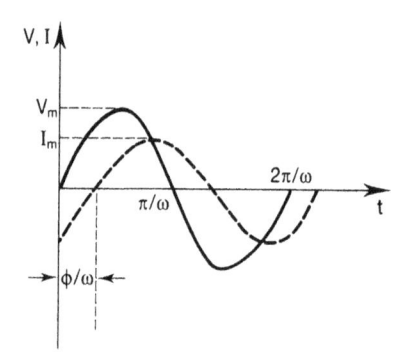

Figure 4.2

Fig. 4.2 shows a sinusoidal voltage waveform of maximum value V_m. This may be represented mathematically by

$$v = V_m \sin \omega t \tag{4.2}$$

where ω is the angular frequency measured in radians per second, related to the frequency f (Hz) by

$$\omega = 2\pi f \tag{4.3}$$

The time axis may be converted into an angle axis simply by multiplying by ω.

Phase difference

The second sinusoidal waveform (shown dashed) in Fig. 4.2 shows a current of maximum value I_m. This waveform is described mathematically by

$$i = I_m \sin(\omega t - \phi) \tag{4.4}$$

and is said to lag the voltage waveform by an angle ϕ because its peak value occurs ϕ/ω seconds after that of the voltage wave. Alternatively we could say that the voltage waveform leads the current waveform by an angle ϕ (i.e. by ϕ/ω seconds). There is said to be a phase difference between the two waveforms.

Example 4.2

Four sinusoidally alternating quantities are represented by:

$a = 5 \sin \omega t$; $b = 15 \sin(\omega t - 30°)$; $c = 10 \sin(\omega t + 60°)$; $d = 5 \sin 2\omega t$

If $\omega = 314$ rad/s:

(1) comment on the relative magnitudes and frequencies of these quantities;

(2) determine the frequency of quantity d;

(3) state the period of quantity b;

(4) state the phase relationship of
   (a) a with respect to b
   (b) a with respect to c
   (c) b with respect to c.

Solution

1 From Equation (4.2) we see that the coefficient of the sine function represents the magnitude of the quantity. Thus the magnitude of d is the same as that of a (5 units); b is three times as big as a (15 units); c is twice as big as a (10 units). From Equation (4.3) we see that $f = \omega/2\pi$ so the frequency of quantities a, b and c is the same at $\omega/2\pi$, whereas that of quantity d is double at $2\omega/2\pi$.

2 The frequency of d is $(2 \times 314)/(2 \times 3.14) = 100$ Hz.

3 The period (T) of quantity b is the reciprocal of its frequency (i.e. $1/f$), which is half that of quantity d at 50 Hz. Therefore $T = 1/50 = 0.02$ s.

4 Taking quantity a as the reference, we see from the sine functions that (a) a leads b by 30°; (b) a lags c by 60°; (c) b lags c by 30° + 60° = 90°.

Phasorial representation of sinusoidal quantities

In Fig. 4.3 the line OP is considered to be rotating in an anticlockwise direction with a constant angular velocity of ω radians per second. Starting from the

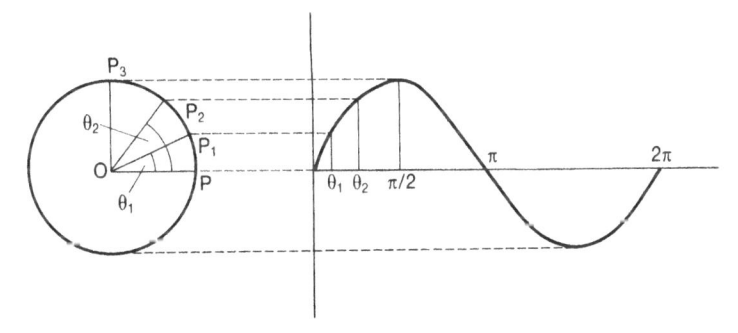

Figure 4.3

horizontal position OP the line will have reached position OP$_1$ after θ_1/ω seconds. After θ_2/ω seconds it will have reached position OP$_2$ and after $\pi/2\omega$ it will be in position OP$_3$. Plotting the horizontal projections of the line as it moves in a circular path results in the sine wave shown. The line OP is called a phasor which is defined as a line whose length represents the magnitude of a sinusoidal quantity and whose position represents its phase with respect to some reference.

Phasor diagrams

Two sinusoidally alternating quantities $v = V_m \sin \omega t$ (a voltage say) and $i = I_m \sin(\omega t + \phi)$, (a current) may be represented by two phasors as shown in

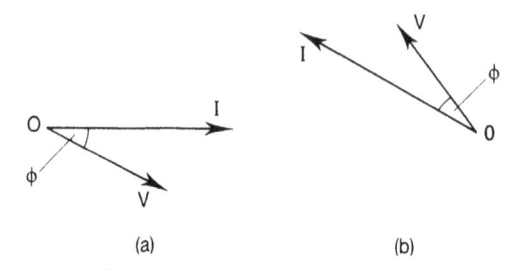

Figure 4.4

Fig. 4.4(a). This is called a phasor diagram and it gives the following information:

- the magnitude of the voltage (the length of the line OV); this can be its maximum value or any constant multiple of it;
- the magnitude of the current (the length of the line OI); again this could be its maximum value;
- the phase difference between the two quantities (the angle ϕ).

The phasors by convention rotate in an anticlockwise direction so that in the example shown the current I leads the voltage V by ϕ. Alternatively we could say that the voltage lags the current by ϕ. Note that the diagram of Fig. 4.4(b) gives exactly the same information as Fig. 4.4(a). The difference between the two is simply that they have been 'stopped' at different instants of time. Only quantities having the same frequency can be represented on the same phasor diagram.

The root mean square (rms) value of a sinusoidal quantity

The rms value of any alternating quantity is found by taking the square root of the mean of the squares of the values of the quantity. It is often called the effective value, and the rms value of a current is that value which has the same heating effect as a steady direct current of the same value. It is represented by a capital letter.

For a sinusoidal current $i = I_m \sin \omega t$. Squaring gives $i^2 = I_m^2 \sin^2 \omega t$ and the mean of this over a complete cycle is $(1/2\pi) \int_0^{2\pi/\omega} I_m^2 \sin^2 \omega t \, d(\omega t)$. The rms value is then the square root of this.

Using the identity $\sin^2 \theta = (1 - \cos 2\theta)/2$,

$$\int_0^{2\pi/\omega} I_m^2 \sin^2 \omega t \, d(\omega t) = I_m^2 \int_0^{2\pi/\omega} (1 - \cos 2\omega t)/2 \, d(\omega t)$$

$$= I_m^2/2 [\omega t - \sin \omega t]_0^{2\pi/\omega} = \pi I_m^2$$

The mean of this (obtained by dividing by 2π) is $I_m^2/2$. The rms value (obtained by taking the square root) is

$$I = I_m/\sqrt{2} \tag{4.5}$$

The average value of a sinusoidal quantity

The average value of a sine wave over a complete cycle is zero, which is rather meaningless, so the average value is taken to be the average over a half cycle. This value is denoted by a capital letter with a subscript (e.g. I_{av}). For a current represented by $i = I_m \sin \omega t$,

$$I_{av} = (1/\pi) \int_0^{2\pi/\omega} I_m \sin \omega t \, d(\omega t) = (I_m/\pi)[-\cos \omega t]_0^{2\pi/\omega}$$

$$= (I_m/\pi)[(-\cos \pi - (-\cos \pi - (-\cos 0)]$$

$$\therefore \quad I_{av} = 2I_m/\pi \tag{4.6}$$

The form factor of an a.c. waveform

This is defined to be the rms value divided by the average value, so that for a sine wave the form factor is $(I_m/\sqrt{2})/(2I_m/\pi) = \pi/(2\sqrt{2})$.

Form factor of a sine wave = 1.11 \hfill (4.7)

Example 4.3

An alternating voltage has an average value of 4 V and a form factor of 1.25. Calculate (1) its rms value, (2) the peak value of a sinusoidal voltage having the same rms value.

Solution

1 The form factor = rms value/average value. Thus the rms value = the form factor × the average value:

$V = 1.25 \times 4 = 5$ V

2 From Equation (4.5) we see that for a sinusoidal voltage the peak value = $\sqrt{2}$ × the rms value. Thus $V_m = \sqrt{2} \times 5 = 7.07$ V.

72 Single-phase a.c. circuits

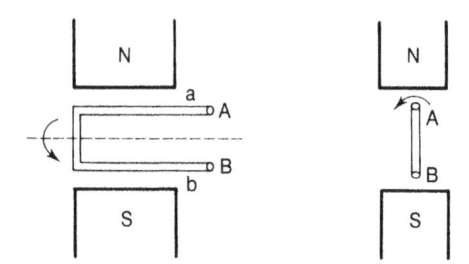

Figure 4.5

Single-phase a.c. quantities

Fig. 4.5 shows an elementary generator. It consists of a single coil having sides a and b rotating in a magnetic field produced by permanent magnets. As the coil sides are cutting magnetic flux, they will have emfs (voltages) induced in them. These emfs could be measured at terminals A and B and would be alternating in nature as the coils come under the influence of first a north pole and then a south pole alternately. Because there are only two terminals and only one emf can be generated in the coil, the a.c. generated is said to be single phase. The emf produced by this simple machine would be alternating but not sinusoidal. By careful machine design, however, generators can be made to generate sinusoidal emfs.

4.2 SINGLE-PHASE A.C. CIRCUITS IN THE STEADY STATE

Steady state operation means that any transient effects following the switching on of a circuit have died away and that the waveforms of voltages and currents are continuous sine waves.

Purely resistive circuits

Fig. 4.6 shows a single-phase voltage source V supplying a pure resistor R. The arrowheads indicate that if the voltage is going positive in the direction shown by its arrowhead then the current will be going positive in the direction shown

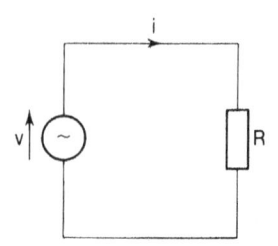

Figure 4.6

by its arrowhead. After the completion of the positive half cycle, of course, both arrowheads will reverse.

Let the voltage be represented by $v = V_m \sin \omega t$. The value of the current i flowing at any instant will be given by $v/R = (V_m/R) \sin \omega t$, i.e. $i = (V_m/R) \sin \omega t$. Now (V_m/R) is the maximum value (I_m) reached by the current, so $i = I_m \sin \omega t$. Note that there is no phase difference between the voltage and the current expressions. The waveforms and the phasor diagram are as shown in Fig. 4.7(a) and (b).

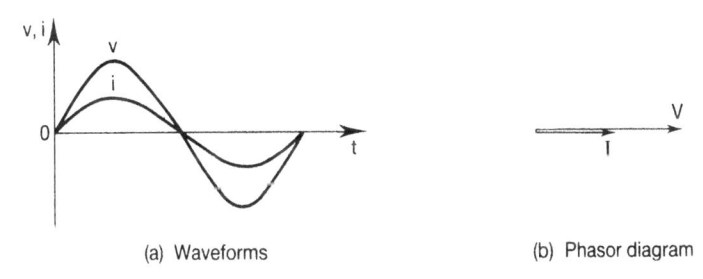

(a) Waveforms (b) Phasor diagram

Figure 4.7

The rms value (V) of the voltage is $V_m/\sqrt{2}$ and that of the current (I) is $I_m/\sqrt{2}$. Now $I_m/\sqrt{2} = (V_m/\sqrt{2})/R$, so that $I = V/R$ (which of course is Ohm's law). For a purely resistive a.c. circuit then,

$$V/I = R \tag{4.8}$$

Example 4.4

In the circuit of Fig. 4.6, $R = 10 \, \Omega$ and $v = 25 \sin 314t$. Determine (1) the rms value of the current, (2) the phase angle of the circuit, (3) the frequency of the supply.

Solution

1. The peak value of the voltage is 25 V so that the peak value of the current is $25/R = 2.5$ A. The rms value of the current is therefore $2.5/\sqrt{2} - 1.77$ A.

2. For a purely resistive circuit the current and voltage phasors are in phase with each other so that the phase angle is zero.

3. The angular frequency is $\omega = 314$ rad s^{-1}. The frequency is
$f = \omega/2\pi = 314/6.28 = 50$ Hz

Purely inductive circuits

The diagram of Fig. 4.8 overleaf shows a pure inductor L connected to a single-phase voltage source V. Let the current be represented by $i = I_m \sin \omega t$. Since this changing current produces a changing flux which will link the inductor then,

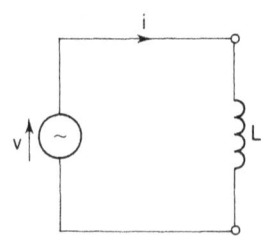

Figure 4.8

according to Faraday's law, an emf will be induced in it, given by $e = -L(di/dt)$. Thus $e = -Ld(I_m \sin \omega t)/dt = -L(I_m \omega \cos \omega t)$, so

$$e = -\omega L I_m \cos \omega t \tag{4.9}$$

The maximum value of this waveform is $\omega L I_m$ and as it is a minus cosine it is $\pi/2$ (90°) behind the current wave. The direction of this emf is such as to oppose the current in accordance with Lenz's law, so the supply voltage V must be equal and opposite to E.

Thus the supply voltage is represented by $v = \omega L I_m \cos \omega t$, which is $\pi/2$ (90°) ahead of the current wave ($\cos \omega t = \sin (\omega t + \pi/2)$). In a purely inductive circuit therefore the current lags the supply voltage by $\pi/2$ (90°). The waveforms and the phasor diagram are shown in Fig. 4.9(a) and (b), respectively.

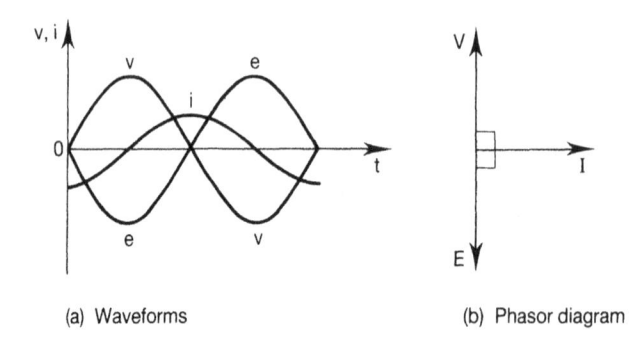

(a) Waveforms   (b) Phasor diagram

Figure 4.9

The maximum value of the voltage wave is $V_m = \omega L I_m$. Dividing V_m and I_m by $\sqrt{2}$ gives $V = \omega L I$, where V and I are now the rms values. Dividing both sides by I we have

$$V/I = \omega L \tag{4.10}$$

Now ωL is written X_L and is called the inductive reactance of the inductor. Its unit is the volt per ampere (the ohm) and because $X_L = \omega L = 2\pi f L$ it varies with frequency. The graphs of Fig. 4.10 show how the inductive reactance and the current vary with frequency in the circuit of Fig. 4.8.

Note that as

$f \to 0$ so $X_L \to 0$ and $I \to \infty$

4.2 Single-phase a.c. circuits in the steady state

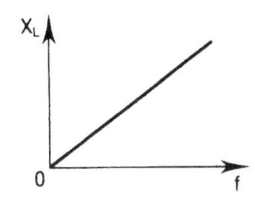

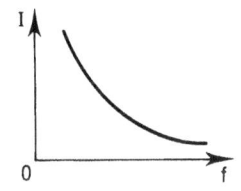

Figure 4.10

and as

$f \to \infty$ so $X_L \to \infty$ and $I \to 0$

Example 4.5

In the circuit of Fig. 4.8, $L = 5$ mH, $i = 10 \sin 2\pi ft$ and $f = 400$ Hz. Calculate (1) the inductive reactance of the circuit, (2) the rms value of the supply voltage.

Solution

1 The inductive reactance, $X_L = 2\pi fL = 2\pi\, 400\, 5 \times 10^{-3} = 12.56\, \Omega$.

2 The rms value of the voltage is given by $V = IX_L$, where I is the rms value of the current. Now

$I = I_m/\sqrt{2} = 10/\sqrt{2} = 7.07$ A

so

$V = 7.07 \times 12.56 = 88.8$ V

Purely capacitive circuits

The diagram of Fig. 4.11 shows a pure capacitor C connected to a voltage source V. We have seen (Chapter 2) that $i = dq/dt$ and that $q = CV$ so that $i = d(Cv)/dt = Cdv/dt$ if C is constant. Let the source voltage be $v = V_m \sin \omega t$. Then

$i = Cdv/dt = Cd(V_m \sin \omega t)/dt = C[V_m \omega \cos \omega t]$

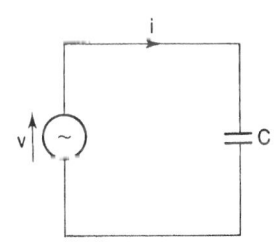

Figure 4.11

i.e.

$$i = \omega C V_m \sin(\omega t + \pi/2)$$

Comparing the current and voltage expressions we see that the current leads the voltage by $\pi/2$ (90°). The waveforms and phasor diagrams are given in Fig. 4.12(a) and (b).

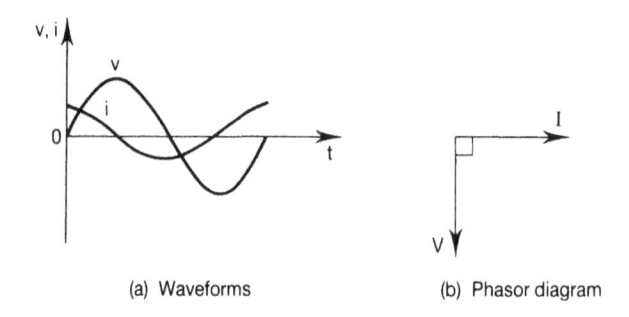

(a) Waveforms (b) Phasor diagram

Figure 4.12

Now $I_m = \omega C V_m$ and dividing both sides by $\sqrt{2}$ gives $I = \omega C V$ (I and V are now rms values). Rearranging we get

$$V/I = 1/\omega C \tag{4.11}$$

Now $1/\omega C$ is written X_C and is called the capacitive reactance of the capacitor, for which the unit is the volt per ampere (the ohm). Since $X_C = 1/\omega C = 1/2\pi f C$ it is dependent upon frequency. Fig. 4.13 shows how the capacitive reactance and the current in the circuit of Fig. 4.11 vary with frequency.

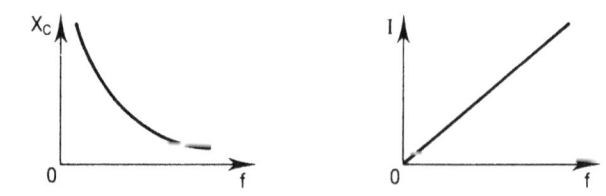

Figure 4.13

Note that as

$f \to 0$ so $X_C \to \infty$ and $I \to 0$

and that as

$f \to \infty$ so $X_C \to 0$ and $I \to \infty$

Example 4.6

In the circuit of Fig. 4.11, $C = 120\,\mu\text{F}$ and the rms values of V and I are, respectively, 100 V and 5 A. Calculate the frequency of the sinusoidal supply voltage.

Solution

The capacitive reactance $X_C = V/I = 100/5 = 20\ \Omega$. From $X_C = 1/2\pi fC$ we see that the frequency $f = 1/2\pi CX_C = 1/(2\pi\ 120 \times 10^{-6} \times 20)$, so $f = 66.3$ Hz.

Summary

For the single-element a.c. circuits, which can also be considered as building blocks for the multiple-element circuits which follow, we have seen that:

For the pure resistor, the current is in phase with the source voltage and
$$V = IR \tag{4.12}$$

For the pure inductor, the current lags the source voltage by 90° ($\pi/2$ radians) and $V = IX_L$ (4.13)

For the pure capacitor, the current leads the source voltage by 90° ($\pi/2$ radians) and $V = IX_C$ (4.14)

The inductive reactance $X_L = 2\pi fL$ (4.15)

The capacitive reactance $X_C = 1/2\pi fC$ (4.16)

4.3 SERIES A.C. CIRCUITS

Series RL circuits

In practice resistive circuits will have some inductance however small because the circuit must contain at least one loop of connecting wire. Also inductive circuits must have some resistance due to the resistance of the wire making up the coil. It is usual to show the resistance of a coil as a separate pure lumped resistor in series with a pure inductor as shown in Fig. 4.14. We can then make use of our building blocks assembled above.

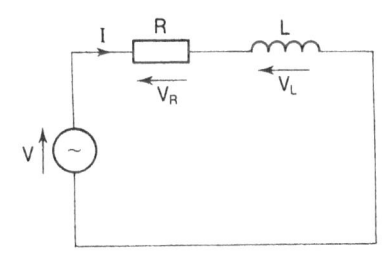

Figure 4.14

Kirchhoff's laws can be applied to a.c. circuits in the same way as for d.c. circuits provided we use phasor sums rather than algebraic sums. Applying

78 Single-phase a.c. circuits

KVL to the circuit of Fig. 4.14 and taking the clockwise direction as being positive,

$V - V_L - V_R = 0$ (phasorially)

so that

$V = V_L + V_R$ (phasorially) $= IX_L + IR$ (phasorially)

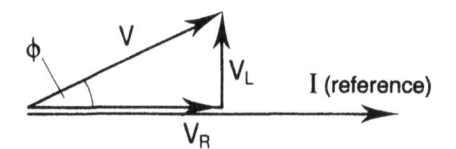

Figure 4.15

This phasor addition is shown in Fig. 4.15. The current is taken to be the reference phasor because it is common to both elements. From Equation (4.12), the voltage drop (IR) in the resistance R is in phase with the current (I). From Equation (4.13), the voltage drop (IX_L) in the inductive reactance X_L is $90°$ ahead of the current. These two voltage drops are then summed to give the source voltage V.

From the phasor diagram we see that, by Pythagoras' theorem, $V^2 = V_R^2 + V_L^2$ so that

$$V^2 = (IR)^2 + (IX_L)^2 = I^2(R^2 + X_L^2)$$

Taking the square root of both sides we get

$$V = I\sqrt{(R^2 + X_L^2)} = IZ \tag{4.17}$$

where $Z = \sqrt{(R^2 + X_L^2)}$ and is called the impedance of the circuit. Since $Z = V/I$ its unit is the volt per ampere (the ohm).

The phasor diagram is now as shown in Fig. 4.16(a) and if we divide all three phasors by I we obtain the diagram of Fig. 4.16(b) which is called an impedance triangle. The angle ϕ is the phase angle of the circuit, and from the geometry of the triangle we see that $R/Z = \cos \phi$, $X_L/Z = \sin \phi$ and $X_L/R = \tan \phi$.

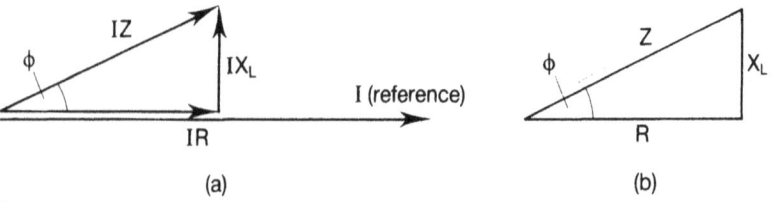

Figure 4.16

Example 4.7

In the circuit of Fig. 4.14, $R = 5 \, \Omega$, $l = 50$ mH and $v = 100 \sin 628t$. Determine (1) the inductive reactance of the circuit, (2) the impedance of the circuit, (3) the current drawn from the supply, (4) the phase angle of the circuit.

Solution

1. From Equation (4.15) the inductive reactance
$X_L = \omega L = 628 \times 50 \times 10^{-3} = 31.4 \ \Omega$

2. From Equation (4.17) the impedance
$Z = \sqrt{(R^2 + X_L^2)} = \sqrt{(5^2 + 31.4^2)} = 31.79 \ \Omega$

3. The current $I = V/Z = 100/31.79 = 3.15$ A.

4. The phase angle $\phi = \tan^{-1}(X_L/R) = \tan^{-1}(31.4/5) = 80.95°$.

Series RC circuits

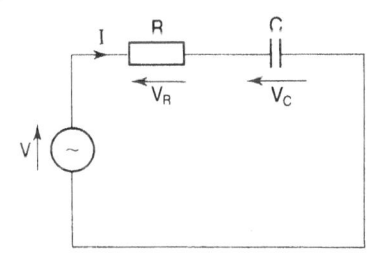

Figure 4.17

Applying KVL to the circuit of Fig. 4.17 and taking the clockwise direction to be positive we have

$V - V_C - V_R = 0$ (phasorially)

so that

$V = V_R + V_C = IR + IX_C$ (phasorially)

With I as the reference phasor, the phasor diagram is shown in Fig. 4.18(a). In

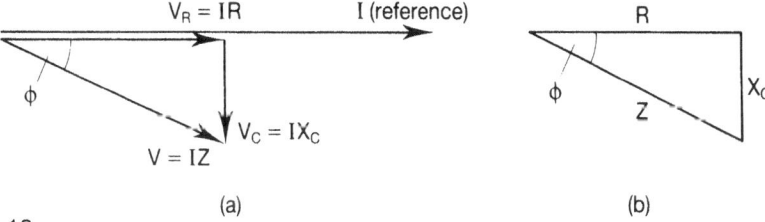

Figure 4.18

this diagram V_R ($=IR$) is in phase with the current (in accordance with Equation (4.12)), and V_C ($=IX_C$) lags the current by 90° (in accordance with Equation (4.14)). Now

$V^2 = (IR)^2 + (IX_C)^2 = I^2(R^2 + X_C^2)$

so that

$V = I\sqrt{(R^2 + X_C^2)} = IZ$

where $Z = \sqrt{(R^2 + X_C^2)}$ and is called the impedance of the circuit.

80 Single-phase a.c. circuits

As in the case of the RL circuit we can divide each of the phasors in Fig. 4.18(a) by I to obtain the impedance triangle for the circuit and this is shown in Fig. 4.18(b). We see that $\cos \phi = R/Z$; $\sin \phi = X_C/Z$ and $\tan \phi = X_C/R$ where ϕ is the phase angle of the circuit.

Example 4.8

In the circuit of Fig. 4.17, $R = 10 \, \Omega$, $C = 10 \, \mu F$ and $f = 400$ Hz. Calculate (1) the impedance of the circuit, (2) the phase angle of the circuit.

Solution

1 The impedance $Z = \sqrt{(R^2 + X_C^2)}$.
 Now $R = 10 \, \Omega$ and $X_C = 1/2\pi f C = 1/(2\pi \, 400 \times 10 \times 10^{-6}) = 39.8 \, \Omega$, so
 $Z = \sqrt{(10^2 + 39.8^2)} = 41 \, \Omega$

2 The phase angle of the circuit is
 $\phi = \tan^{-1}(X_C/R) = \tan^{-1}(39.8/10) = 75.9°$

Series RLC circuits

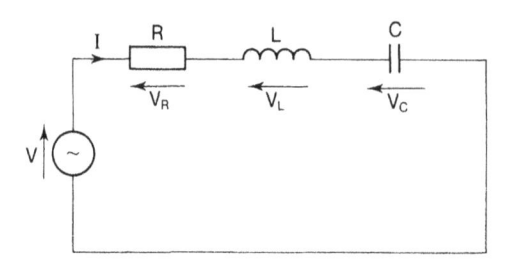

Figure 4.19

Applying KVL to the circuit of Fig. 4.19 and taking the clockwise direction to be positive,

$V - V_R - V_L - V_C = 0$ (phasorially)
$V = V_R + V_L + V_C = IR + IX_L + IX_C$ (phasorially)

The phasor diagram is drawn in Fig. 4.20(a) assuming that $X_L > X_C$ and in Fig. 4.20(b) assuming that $X_L < X_C$. In both cases the voltage V_R ($=IR$) is drawn in phase with the reference phasor (I), the voltage V_L ($=IX_L$) is drawn 90° ahead of the current I and the voltage V_C is drawn 90° behind the current I in accordance with Equations (4.12), (4.13) and (4.14), respectively.

In the first case (Fig. 4.20(a)) we see that the current lags the voltage V by the phase angle ϕ and that therefore the circuit behaves as an inductive circuit. Also we have that

$V^2 = (IR)^2 + [I(X_L - X_C)]^2$

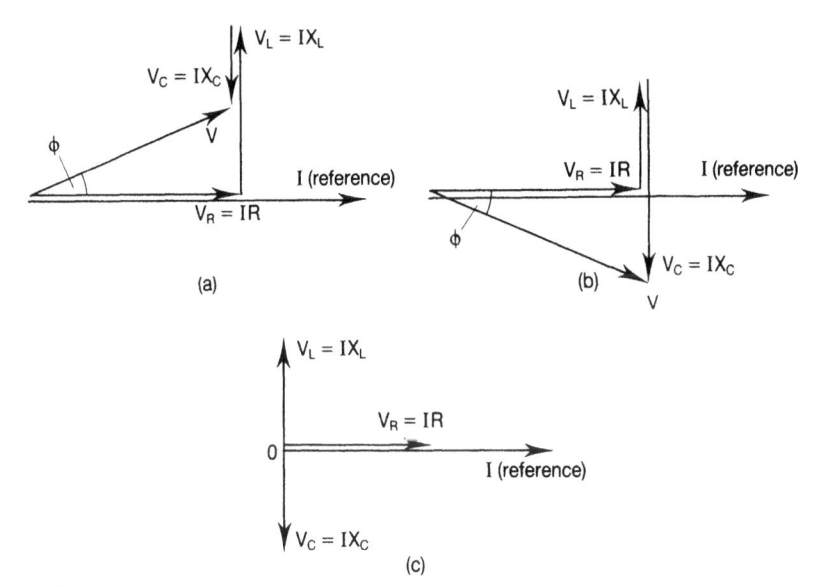

Figure 4.20

and

$$V = I\sqrt{(R^2 + [X_L - X_C]^2)}$$

The impedance is

$$Z\ (=V/I) = \sqrt{(R^2 + [X_L - X_C]^2)}$$

The phase angle is obtained from $\cos \phi = R/Z$ or $\sin \phi = (X_L - X_C)/Z$ or $\tan \phi = (X_L - X_C)/R$.

In the second case with $X_C > X_L$ we see that the current leads the voltage V by the phase angle ϕ and that therefore the circuit behaves as a capacitive circuit. Also we have that $V^2 = (IR)^2 + (I[X_C - X_L])^2$ and so

$$V = I\sqrt{(R^2 + [X_C - X_L]^2)} \tag{4.18}$$

The impedance is $Z = \sqrt{(R^2 + [X_C - X_L]^2)}$ and the phase angle is $\phi = \tan^{-1}(X_C - X_L)/R$.

There is a third possibility for this circuit: that the inductive reactance X_L is equal to the capacitive reactance X_C. In this case the phasor diagram takes the form shown in Fig. 4.20(c), from which we see that $V = IR$, confirmed by putting $X_L = X_C$ in Equation (4.18). The circuit then behaves as a purely resistive circuit and is a special case which is fully discussed in Chapter 6.

Example 4.9

In the circuit of Fig. 4.19, $R = 12\ \Omega$, $L = 150$ mH, $C = 10\ \mu\text{F}$, $v = 100 \sin 2\pi ft$ and $f = 100$ Hz. Calculate (1) the impedance of the circuit, (2) the current drawn from the supply, (3) the phase angle of the circuit. Give an expression from which the current at any instant could be determined.

82 Single-phase a.c. circuits

Solution

1 The impedance $Z = \sqrt{(R^2 + [X_L - X_C]^2)}$. Now

$$X_L = 2\pi fL = 2\pi\, 100 \times 150 \times 10^{-3} = 94.2\,\Omega$$

and

$$X_C = 1/2\pi fC = 1/2\pi\, 100 \times 10 \times 10^{-6} = 159\,\Omega$$

so

$$Z = \sqrt{(12^2 + [94.2 - 159]^2)} = 65.9\,\Omega$$

2 The current drawn from the supply is given by $I = V/Z$ where I and V are rms values. The peak value of the supply voltage is 100 V so that the rms value is $100/\sqrt{2} = 70.7$ V, so

$$I = V/Z = 70.7/65.9 = 1.07\,\text{A}$$

3 The phase angle of the circuit is $\phi = \cos^{-1}(R/Z) = \cos^{-1}(12/65.9) = 79.5°$. Because the capacitive reactance is greater than the inductive reactance, the circuit is predominantly capacitive and so the phase angle is a leading one (i.e. the current leads the voltage).

The expression for the current is $i = I_m \sin(2\pi ft - \phi)$ with $I_m = \sqrt{2}I$ and $\phi = 79.5° = 79.5\pi/180$ rad. Thus $i = 1.5\sin(200\pi t - 0.44\pi)$ A.

4.4 COMPLEX NOTATION

We have seen that a phasor quantity is one for which both magnitude and direction is important. These quantities may be represented by phasor diagrams in the manner shown earlier in the chapter. In Fig. 4.21(a), the phasor V_1 is

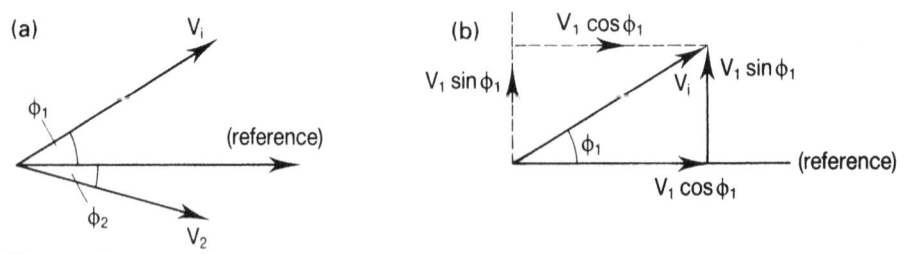

Figure 4.21

shown to be leading the reference phasor by ϕ_1 degrees, whereas the phasor V_2 is shown as lagging the reference by ϕ_2 degrees. It is conventional to take the horizontal axis as the reference direction.

V_1 may be represented by

$$V_1 = |V_1|\angle\phi_1$$

In this notation $|V_1|$ indicates the magnitude of the quantity and is represented by the length of the phasor, while $\angle \phi_1$ indicates that it is ϕ_1 degrees ahead of (leading) the reference direction. Similarly,

$$V_2 = |V_2| \angle -\phi_2$$

the minus sign indicating that V_2 is ϕ_2 degrees behind (lagging) the reference direction.

Note that V_1 has two components at right angles as shown in Fig. 4.21(b). These are V_{1r} along the reference direction and V_{1q} at right angles to the reference direction. Now $V_{1r} = V_1 \cos \phi_1$ and $V_{1q} = V_1 \sin \phi_1$ so that

$$V_1 = V_{1r} + V_{1q} = V_1 \cos \phi_1 + V_1 \sin \phi_1 \quad \text{(a phasorial addition)}$$

To indicate that $V_1 \sin \phi_1$ is at right angles to $V_1 \cos \phi_1$ an operator j is introduced which, when placed before a quantity simply indicates that that quantity has been shifted through 90° in an anticlockwise direction with respect to any quantity which does not have a j in front of it. (Incidentally mathematicians use the letter i placed after the quantity.)

Using this notation we have that

$$V_1 = V_1 \cos \phi_1 + jV_1 \sin \phi_1 = V_1[\cos \phi_1 + j \sin \phi_1]$$

and

$$V_2 = V_2 \cos \phi_2 - jV_2 \sin \phi_2 = V_2[\cos \phi_2 - j \sin \phi_2]$$

The minus sign in front of the j indicates that the quantity $V_2 \sin \phi_2$ has been shifted through 90° in a *clockwise* direction.

The Argand diagram

In Fig. 4.22(a), which is called an Argand diagram, the phasor V_1 is shown in the positive reference direction. The phasor V_2 has the same length as V_1 but is 90° ahead of it so, using the complex (or 'j') notation, we write

$$V_2 = jV_1 \tag{4.19}$$

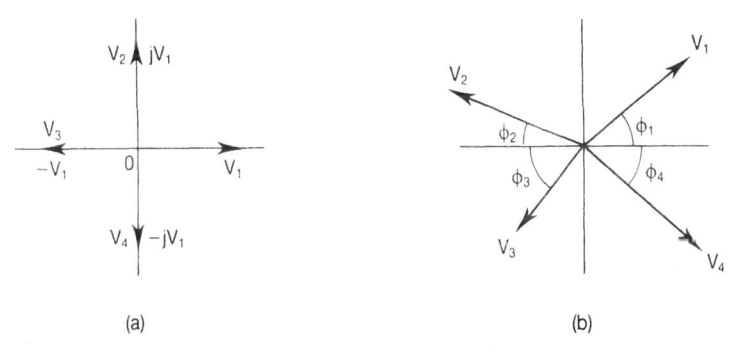

Figure 4.22

The phasor V_3 has the same length as V_2 but is 90° ahead of it so that $V_3 = jV_2$. But from equation (4.19) $V_2 = jV_1$, so

$$V_3 = j(jV_1) = j^2V_1 \tag{4.20}$$

Now V_3 has the same length as V_1 and is in the opposite direction to it, which means that $V_3 = -V_1$. It follows that

$$j^2 = -1 \quad \text{and} \quad j = \sqrt{-1} \tag{4.21}$$

Since it is impossible to take the square root of -1, it ($\sqrt{-1}$) is said to be imaginary and the vertical or j-axis is often referred to as the imaginary axis. It is also known as the quadrature axis. The horizontal or reference axis is also called the real axis.

Now V_4 has the same length as V_3 and is 90° ahead of it so that $V_4 = jV_3$. From equation (4.20) $V_3 = j^2V_1$, so

$$V_4 = j(j^2V_1) = j(-1)V_1 = -jV_1$$

and lies along the negative imaginary axis. Finally, applying the j operator to V_4 shifts it through 90° in an anticlockwise direction bringing it to the positive real axis as V_1. This is verified by noting that

$$jV_4 = j(-jV_1) = -j^2V_1 = -(-1)V_1 = V_1$$

Note also that

$$jV_4 = j(jV_3) = jj(jV_2) = jjj(jV_1) = j^4V_1 = V_1$$

It follows that

$$j^4 = 1 \tag{4.22}$$

The Argand diagram consists of four quadrants.

- In the first quadrant, the real and imaginary axes are both positive and, as shown in Fig. 4.22(b), the angle ϕ_1 takes values between 0 and 90° from the real positive direction.

- In the second quadrant, the real axis is negative and the imaginary axis is positive and phasors lying in this quadrant are between 90° and 180° from the reference direction.

- In the third quadrant, both the real and the imaginary axes are negative and phasors are between 180° and 270° from the reference.

- Finally, in the fourth quadrant the real axis is positive and the imaginary axis is negative, the angles from the reference direction being between 270° and 360°.

Note that in all quadrants the angles (ϕ_1, ϕ_2, ϕ_3, and ϕ_4) are obtained from $\tan^{-1}$ (imaginary component/real component).

Example 4.10

Represent the following currents on an Argand diagram: (1) $I_1 = (2 + j3)$ A
(2) $I_2 = (-5 + j2)$ A (3) $I_3 = (4 - j4)$ A (4) $I_4 = (-4 - j5)$ A
(5) $I_5 = 5 \angle 100°$ A (6) $I_6 = 3 \angle -10°$ A.

Solution

1. I_1 has 2 units along the positive real axis and 3 units along the positive imaginary axis. Its length is $\sqrt{(2^2 + 3^2)} = 3.61$ A.

2. I_2 has 5 units along the negative real axis and 2 units along the positive imaginary axis. Its magnitude is $\sqrt{(5^2 + 2^2)} = 5.39$ A.

3. I_3 has 4 units along the positive real axis and 4 units along the negative imaginary axis. Its magnitude is $\sqrt{(4^2 + 4^2)} = 5.66$ A.

4. I_4 has 4 units along the negative real axis and 5 units along the negative imaginary axis. Its magnitude is $\sqrt{(4^2 + 5^2)} = 6.4$ A.

5. I_5 has a magnitude of 5 A and is at 100° in an anticlockwise (positive) direction from the reference (i.e. the positive real axis).

6. I_6 has a magnitude of 3 A and is placed 10° in a clockwise (negative) direction from the positive real axis.

These are shown on the Argand diagram in Fig. 4.23.

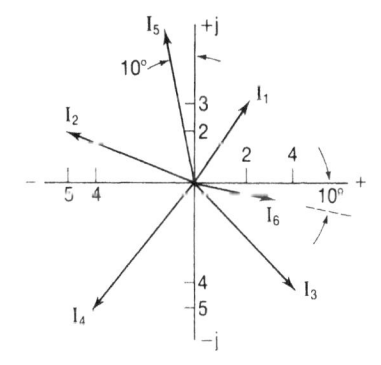

Figure 4.23

Rectangular and polar coordinates

The phasor diagram of Fig. 4.24 shows a phasor V and its two components in

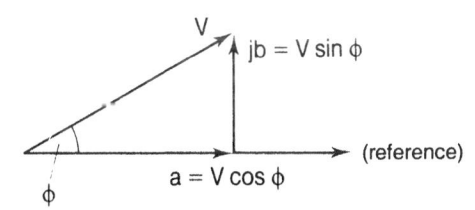

Figure 4.24

86 Single-phase a.c. circuits

the real and imaginary axes. Note that $V = |V|\angle\phi = V\cos\phi + jV\sin\phi = a + jb$ say (with $a = V\cos\phi$ and $b = V\sin\phi$). From the geometry of the diagram we see that $V = \sqrt{(a^2 + b^2)}$ and that $\phi = \tan^{-1}(b/a)$. Thus

$$V = |V|\angle\phi = \sqrt{(a^2 + b^2)}\angle\tan^{-1}(b/a) \tag{4.23}$$

This is called the polar coordinate form of the phasor V. Also

$$V = |V|\cos\phi + j|V|\sin\phi = a + jb \tag{4.24}$$

This is called the rectangular coordinate form of the phasor V. It is a simple matter to change from one form to the other.

Example 4.11

Express (1) $I = (4 + j3)$ A in polar form, (2) $V = 25\angle-30°$ V in rectangular form.

Solution

1 The magnitude of the current is $\sqrt{(4^2 + 3^2)} = 5$ A.
  The angle $\phi = \tan^{-1}(3/4) = 36.86°$.
  Thus in polar form we have $I = 5\angle 36.86°$. The current is shown in both forms in Fig. 4.25(a).

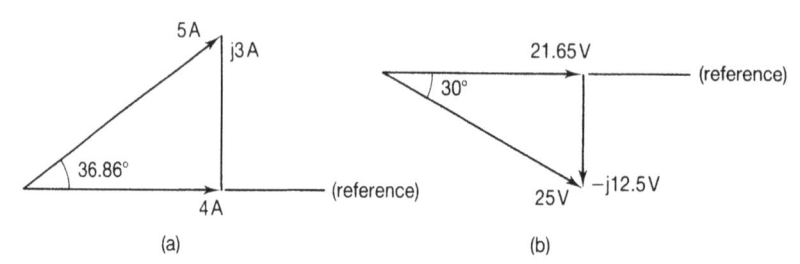

Figure 4.25

2 The component of V along the real axis is $25\cos 30° = 21.65$ V.
  The component of V along the negative imaginary axis is
  $25\sin 30° = 12.5$ V.
  Thus in rectangular form we have $V = (21.65 - j12.5)$ V. The voltage is shown in both forms in Fig. 4.25(b).

Addition and subtraction of complex quantities

It is more convenient to do addition and subtraction using the rectangular coordinates form of the quantities. The real parts of the quantities are added (or subtracted) to give the real part of the resultant quantity. Similarly the imaginary parts are added (or subtracted) to give the imaginary part of the resultant.

Example 4.12

Determine the sum of the two voltages $V_1 = (10 + j50)$ V and $V_2 = (15 - j25)$ V.

Solution

Let the resultant voltage be V. Then

$$V = (10 + 15) + j(50 - 25) = (25 + j25) \text{ V}$$

The magnitude of V is given by $\sqrt{(25^2 + 25^2)} = 35.35$ V. The phase of V with respect to the reference is given by $\phi = \tan^{-1}(25/25) = 45°$. The complete phasor diagram is shown in Fig. 4.26.

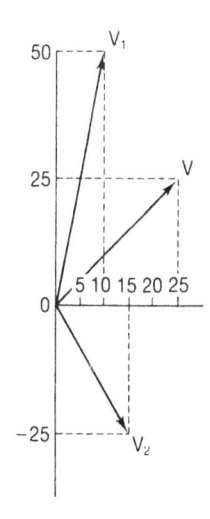

Figure 4.26

Example 4.13

Two currents entering a node in a circuit are given by $I_1 = 20\angle 30°$ A and $I_2 = 30\angle 45°$ A. Calculate the magnitude and phase of a third current I_3 leaving the node.

Solution

Converting to rectangular coordinates we have

$I_1 = 20 \cos 30° + j20 \sin 30° = (17.32 + j10)$ A
$I_2 = 30 \cos 45° + j30 \sin 45° = (21.2 + j21.2)$ A

By Kirchhoff's current law,

$I_3 = I_1 + I_2 = (17.32 + 21.2) + j(10 + 21.2) = (38.52 + j31.2)$ A
$I_3 = \sqrt{(38.52^2 + 31.2^2)} \angle \tan^{-1}(31.2/38.52) = 49.57\angle 39°$ A

Example 4.14

The voltages at two points A and B are given by $V_A = 100 \angle 20°$ V and $V_B = 200 \angle -25°$ V. Calculate the potential difference $(V_A - V_B)$ between them.

Solution

Subtraction is best carried out using rectangular coordinates, so converting to this form we have

$V_A = 100 \cos 20° + j100 \sin 20° = (93.9 + j34.2)$ V
$V_B = 200 \cos 25° - j200 \sin 25° = (181 - j84.5)$ V

$V_A - V_B = (93.9 - 181) + j(34.2 + 84.5) = (-87.1 + j118.7)$ V
$ = 147.2 \angle 126.3°$ V

Example 4.15

Subtract $I_1 = (10 + j5)$ A from $I_2 = (5 - j15)$ A.

Solution

$I_2 - I_1 = (5 - j15) - (10 + j5)$
$ = 5 - j15 - 10 - j5$
$ = (-5 - j20)$ A

Multiplication and division of complex quantities

These operations are best carried out using polar coordinates. For multiplication the magnitudes are multiplied and the angles are added; for division the magnitudes are divided and the angles are subtracted.

Example 4.16

If $A = (5 + j6)$ and $B = (7 - j10)$, obtain the product AB.

Solution

First let us use rectangular coordinates.

$A \times B = (5 + j6)(7 - j10) = (5 \times 7) - j(5 \times 10) + j(6 \times 7) - j^2(6 \times 10)$
$ = 35 - j50 + j42 + 60$
$ = 95 - j8$

In polar form

$A \times B = \sqrt{(95^2 + 8^2)} \angle -\tan^{-1}(8/95) = 95.33 \angle -4.8°$

In polar form

$A = \sqrt{(5^2 + 6^2)} \angle \tan^{-1}(6/5) = 7.81 \angle 50.2°$ V

and

$B = \sqrt{(7^2 + 10^2)} \angle -\tan^{-1}(10/7) = 12.21 \angle -55°$ A

$A \times B = AB \angle (\phi_A + \phi_B) = 7.81 \times 12.21 \angle (50.2 - 55)$

$\quad = 95.33 \angle -4.8°$

Example 4.17

Divide $A = (5 + j6)$ by $B = (7 - j10)$.

Solution

Using rectangular form we have $A/B = (5 + j6)/(7 - j10)$. To proceed from here we have to 'rationalize' the denominator, i.e. remove the j. This is done by multiplying the numerator and the denominator by the conjugate of the denominator. The conjugate of a complex number $(a + jb)$ is $(a - jb)$, obtained simply by changing the sign of the j term. When multiplying a complex number by its conjugate the j disappears. Thus

$(7 - j10)(7 + j10) = (7 \times 7) + (7 \times j10) - j(10 \times 7) - j^2(10 \times 10)$
$\qquad = 49 + j70 - j70 - (-1)(100) = 49 + 100$
$\qquad = 149$

Hence

$A/B = [(5 + j6)(7 + j10)]/[(7 - j10)(7 + j10)]$
$\quad = (35 + j50 + j42 + j^2 60)/149$
$\quad = (-25 + j92)/149$
$\quad = (-0.168 + j0.617)$

Converting to polar form we have $A/B = \sqrt{(0.168^2 + 0.617^2)} \angle \tan^{-1}(0.617/0.168) = 0.64 \angle 74.8°$. Since the real part is negative and the imaginary part is positive then the phasor falls in the second quadrant so that the angle is $180 - 74.8 = 105.2°$ from the real positive axis reference.

From Example 4.16 we have that in polar form these two quantities are

$A = 7.81 \angle 50.2°$ and $B = 12.21 \angle -55°$

Now $A/B = A/B \angle (\phi_A - \phi_B) = 7.81/12.21 \angle [50.2 - (-55)] = 0.64 \angle 105.2°$

Application to the analysis of series a.c. circuits

The phasor diagram for the series circuit of Fig. 4.14 is given in Fig. 4.15. Note that the phasor V_R lies along the reference axis and that the phasor V_L is 90°

ahead of the reference axis. The total circuit voltage V is the phasor sum of these two. In complex form we write:

$$V = V_R + jV_L$$

the j in front of V_L indicating that it is 90° ahead of the reference (and V_R). Thus

$$V = IR + jIX_L = I(R + jX_L) = IZ$$

where Z is the impedance of the circuit:

$$Z = R + jX_L \tag{4.25}$$

Similarly, for the series RC circuit of Fig. 4.17 and its phasor diagram of Fig. 4.18(a)

$$V = V_R - jV_C$$

the $-j$ in front of V_C indicating that it is 90° *behind* the reference. Thus

$$V = I(R - jX_C) = IZ$$

where again Z is the impedance of the circuit:

$$Z = R - jX_C \tag{4.26}$$

For the series RLC circuit of Fig. 4.19 and its associated phasor diagram (Fig. 4.20(a))

$$V = V_R + j(V_L - V_C)$$

Thus

$$V = I[R + j(X_L - X_C)] = IZ$$

$$Z = R + j(X_L - X_C) \tag{4.27}$$

If $X_L > X_C$ the j term is positive indicating a predominantly inductive circuit. For $X_L < X_C$ the j term is negative, indicating a predominantly capacitive circuit. When $X_L = X_C$ there is no j term and the circuit is purely resistive.

The impedance triangles of Fig. 4.16(b) and 4.18(b) take the forms given in Fig. 4.27(a) and (b), respectively.

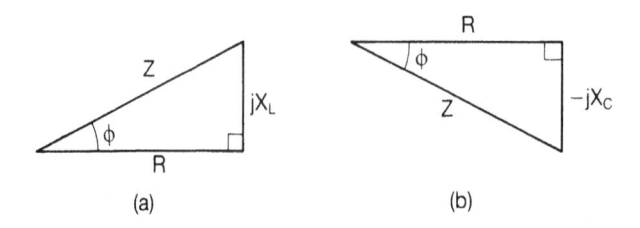

Figure 4.27

Example 4.18

A coil having a resistance of 2.5 Ω and the inductance of 60 mH is connected in series with a capacitor having a capacitance of 6.8 μF to a 230 V, 50 Hz supply. Determine the current drawn from the supply

Solution

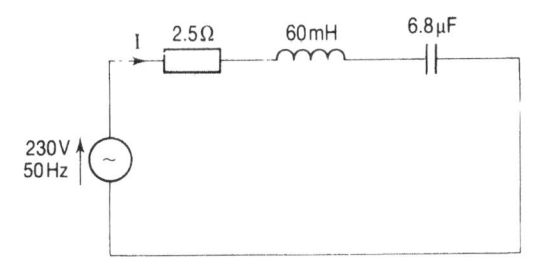

Figure 4.28

The circuit is shown in Fig. 4.28. Take the voltage as the reference so that $V = 230\angle 0°$ V. The impedance of the circuit is given by $Z = R + j(X_L - X_C)$. Now

$$X_L = 2\pi f L = 2\pi 50 \times 60 \times 10^{-3} = 18.85 \text{ }\Omega$$

and

$$X_C = 1/2\pi f C = 1/(2\pi 50 \times 6.8 \times 10^{-6}) = 468 \text{ }\Omega$$

Thus $X_L - X_C = -449.15$ Ω and $Z = (2.5 - j\,449.15)$ Ω. In polar form

$$Z = \sqrt{(2.5^2 + 449.15^2)} \angle \tan^{-1}(449.15/2.5) = 449.2 \angle -89.42° \text{ }\Omega$$

The current is

$$I = V/Z = 230\angle 0°/449.2\angle -89.42° = 230/449.2 \angle [0° - (-89.42°)]$$

so $I = 0.512\angle 89.42°$ A. The current therefore leads the voltage as expected in a predominantly capacitive circuit.

4.5 PARALLEL A.C. CIRCUITS

Parallel RL circuits

The circuit on the following page of Fig. 4.29(a) shows a resistor R in parallel with an inductor L. The corresponding phasor diagram is given in Fig. 4.29(b). It is convenient to take the voltage V as the reference phasor since it is common to both elements.

The current I_R is in phase with the voltage and the current I_L is 90° behind (lagging) the voltage. Kirchhoff's current law tells us that the total current I is the phasor sum of I_R and I_L. The magnitude of I is $\sqrt{(I_R^2 + I_L^2)}$ and the phase angle ϕ is given by $\tan^{-1}(I_L/I_R)$. Since $I = V/Z$, $I_R = V/R$ and $I_L = V/X_L$

$$V/Z = \sqrt{[(V/R)^2 + (V/X_L)^2]} = V\sqrt{[(1/R)^2 + (1/X_L)^2]}$$

92 Single-phase a.c. circuits

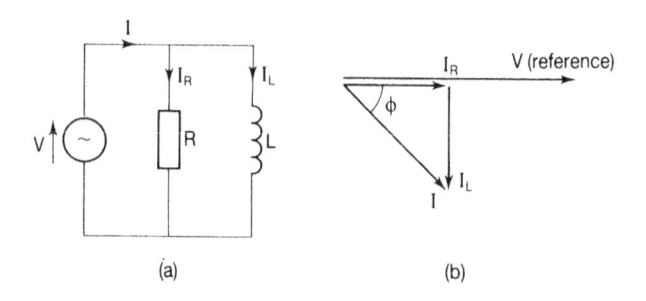

Figure 4.29

Dividing through by V we see that

$$1/Z = \sqrt{[(1/R)^2 + (1/X_L)^2]} \tag{4.28}$$

We have seen (Chapter 2) that the reciprocal of resistance $(1/R)$ is called conductance (G). The reciprocal of reactance $(1/X)$ is called susceptance (B) and the reciprocal of impedance $(1/Z)$ is called admittance (Y), so that Equation (4.28) may be rewritten as

$$Y = \sqrt{(G^2 + B_L^2)} \tag{4.29}$$

In complex form we have the following relationship:

$$I = I_R - jI_L$$

so that

$$V/Z = V/R - jV/X_L$$

Dividing throughout by V gives

$$1/Z = 1/R - j1/X_L$$

and

$$Y = G - jB_L \tag{4.30}$$

If we divide each phasor in Fig. 4.29(b) by V we obtain the admittance triangle shown in Fig. 4.30.

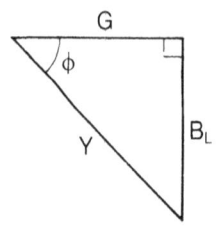

Figure 4.30

Note that the phase angle ϕ is given by $\tan^{-1}(B_L/G) = \tan^{-1}(R/X_L)$ and since $X_L = \omega L$, we have

$$\phi = \tan^{-1}(R/\omega L) \tag{4.31}$$

Example 4.19

Determine the magnitude and phase of the current drawn from the supply in the circuit of Fig. 4.31.

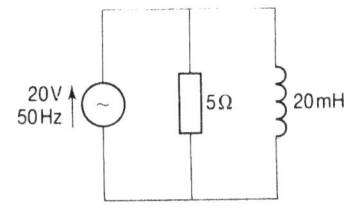

Figure 4.31

Solution

From Equation (4.30) the admittance of the circuit is given by $Y = G - jB_L$. Now $G = 1/R = 1/5 = 0.2$ S, and

$$B_L = 1/X_L = 1/2\pi fL = 1/(2\pi\, 50 \times 20 \times 10^{-3}) = 0.16 \text{ S}$$

Therefore

$$Y = (0.2 - j\,0.16) \text{ S} = 0.26\angle -38.7° \text{ S}$$

Let the voltage be the reference so that $V = 20\angle 0°$ V. Then the current $I = V/Z$ and since $Z = 1/Y$

$$I = VY \tag{4.32}$$

Therefore

$$I = 20 \times 0.26\angle[0° + (-38.7°)] = 5.2\angle -38.7° \text{ A}$$

The current therefore lags the voltage as expected in an inductive circuit.

Parallel RC circuits

The diagram of Fig. 4.32(a) shows a resistor R in parallel with a capacitor C. The corresponding phasor diagram is given in Fig. 4.32(b). In this case, with the voltage V as the reference phasor, we have I_R in phase with V and I_C leading V by 90°. From KCL we have that $I = I_R + I_C$ phasorially.

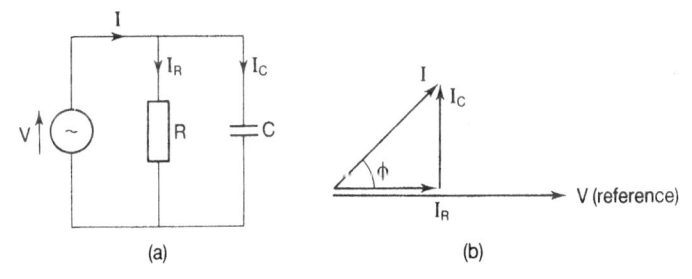

Figure 4.32

The magnitude of I is thus $\sqrt{(I_R^2 + I_C^2)}$ and the phase angle ϕ is $\tan^{-1}(I_C/I_R)$. Since $I = V/Z$, $I_R = V/R$ and $I_C = V/X_C$ we have

$$V/Z = \sqrt{[(V/R)^2 + (V/X_C)^2]} = V\sqrt{[(1/R)^2 + (1/X_C)^2]} = V\sqrt{(G^2 + B_C^2)} = VY$$

In complex form we have

$$I(=V/Z) = I_R + jI_C = V/R + jV/X_C$$

so that

$$1/Z = 1/R + j1/X_C$$

and

$$Y = G + jB_C \tag{4.33}$$

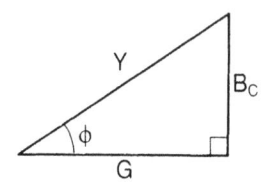

Figure 4.33

From the admittance triangle shown in Fig. 4.33 we see that the phase angle ϕ is given by $\tan^{-1}(B_C/G) = \tan^{-1}(R/X_C)$ and since $X_C = 1/\omega C$, we have

$$\phi = \tan^{-1}(\omega CR) \tag{4.34}$$

Example 4.20

A circuit consisting of a resistor of resistance 5 Ω in parallel with a capacitor of 10 μF capacitance is fed from a 24 V, 4 kHz supply. Calculate the current in magnitude and phase.

Solution

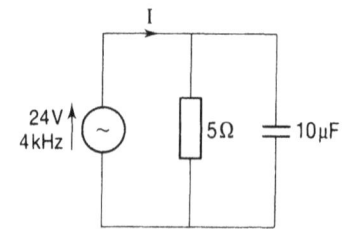

Figure 4.34

The circuit is shown in Fig. 4.34. From Equation (4.33) we have that the admittance is $Y = G + jB_C$. Now

$$G = 1/R = 1/5 = 0.2 \text{ S}$$

and

$$B_C = 1/X_C = 1/(1/2\pi fC) = 2\pi fC = 2\pi \times 4 \times 10^3 \times 10 \times 10^{-6} = 0.25 \text{ S}$$

so that

$$Y = (0.2 + j\,0.25) \text{ S} = 0.32\angle 51.34° \text{ S}$$

Let the voltage be the reference so that $V = 24\angle 0°$ V. Then the current $I = VY = 24 \times 0.32\angle(0 + 51.34) = 7.68\angle 51.34°$ A. This indicates that the current leads the voltage as is to be expected in a capacitive circuit.

4.6 SERIES–PARALLEL A.C. CIRCUITS

The circuit of Fig. 4.35(a) consists of a capacitor C in parallel with an inductor L and a resistor R in series. The phasor diagram is shown in Fig. 4.35(b).

To draw the phasor diagram we first choose V to be the reference since this is common to both branches. The current I_C through the capacitor will be 90°

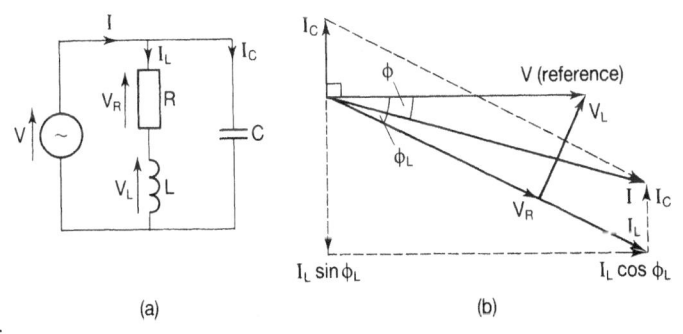

Figure 4.35

ahead of V. The current I_L through the inductor and resistor will be ϕ_L behind V where $\phi_L = \tan^{-1}(\omega L/R)$. The current I_L is now taken as a reference for the RL branch.

We can draw the voltage drop $V_R (= I_L R)$ in phase with the current I_L which produces it and the voltage $V_L (= I_L \omega L)$ 90° ahead of the current I_L which produces it. The phasor sum of V_R and V_L must be V, the total circuit voltage, and the phasor sum of I_C and I_L is I, the total circuit current.

The phase angle of the circuit is ϕ and is shown to be lagging V. In practice of course whether the total current is leading or lagging will depend upon the relative magnitudes of I_C and the quadrature component of I_L (i.e. $I_L \sin \phi_L$).

If $I_C > I_L \sin \phi_L$ then I will lead V and the circuit is predominantly capacitive.

If $I_C < I_L \sin \phi_L$ then I will lag V and the circuit is predominantly inductive.

If $I_C = I_L \sin \phi_L$ then I is in phase with V and the circuit behaves as a pure resistor.

From the geometry of the phasor diagram

96 *Single-phase a.c. circuits*

$$I^2 = (I_L \cos \phi_L)^2 + (I_L \sin \phi_L - I_C)^2 \tag{4.35}$$

and

$$\phi = \tan^{-1} [(I_L \sin \phi_L - I_C)/I_L \cos \phi_L \tag{4.36}$$

The case where ϕ turns out to be 0, i.e. when $I_L \sin \phi_L = I_C$ is a special case which will be fully discussed in Chapter 6.

In complex form, we have, from the phasor diagram of Fig. 4.35(b),

$$I = I_L \cos \phi_L + j(I_C - I_L \sin \phi_L) \tag{4.37}$$

If $I_C > I_L \sin \phi_L$ then the j term is positive indicating that the total current leads the voltage and that the circuit is therefore predominantly capacitive.

If $I_C < I_L \sin \phi_L$ then the j term is negative indicating that the total current lags the voltage and that the circuit is therefore predominantly inductive.

Also we have that $V = V_C = V_R + V_L$ phasorially.

Example 4.21

Determine (1) the total current drawn from the supply in the circuit of Fig. 4.36, (2) the phase angle of the inductive branch and of the circuit as a whole.

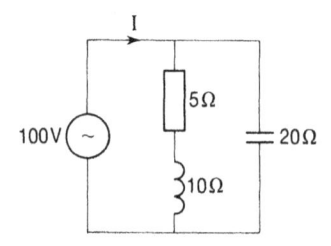

Figure 4.36

Solution

1 The current through the capacitor is $I_C = V/X_C = 100/20 = 5$ A. The current through the inductive branch is $I_L = V/Z_L$ where $Z_L = R + jX_L$. The phase angle of the inductive branch is given by

$\phi_L = \tan^{-1}(X_L/R) = \tan^{-1}(10/5) = 63.4°$

$Z_L = \sqrt{(R^2 + X_L^2)} = \sqrt{(5^2 + 10^2)} = 11.18 \, \Omega$

$I_L = V/Z_L = 100/11.18 = 8.9$ A

From Equation (4.35)

$I^2 = (I_L \cos \phi_L)^2 + (I_L \sin \phi_L - I_C)^2$
$= (8.9 \cos 63.4)^2 + (8.9 \sin 63.4 - 5)^2 = 15.9 + 8.75 = 24.65 \, A^2$

Therefore

$I = 4.96$ A

2 We have already calculated the phase angle of the inductive branch to be 63.4° and this is of course lagging. The phase angle of the circuit as a whole is given by Equation (4.36) to be

$$\phi = \tan^{-1}[(I_L \sin \phi_L - I_C)/(I_L \cos \phi_L)] = \tan^{-1}(2.96/3.99) = 36.5°$$

Because $I_L \sin_L > I_C$ then the circuit is predominantly inductive and so the phase angle is lagging.

4.7 POWER IN SINGLE-PHASE A.C. CIRCUITS

We saw in Chapter 3 that in d.c. circuits, power (P) is the product of voltage (V) and current (I):

$$P = VI \quad \text{watts} \tag{4.38}$$

In a.c. circuits, where the voltage and current are both changing from instant to instant, the instantaneous power (p) is the product of the instantaneous voltage (v) and the instantaneous current (i) i.e. $p = vi$.

Purely resistive circuits

For the purely resistive circuit shown in Fig. 4.6 the waveforms of voltage, current and power are given in Fig. 4.37. Note that the power waveform never goes negative (the product vi is always positive) and that its frequency is twice that of the voltage and current waveforms.

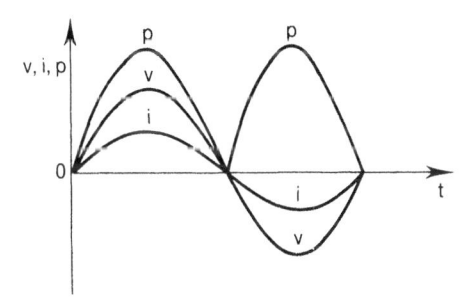

Figure 4.37

If $v = V_m \sin \omega t$ then $i = I_m \sin \omega t$ and since $p = vi$,

$$p = V_m I_m \sin^2 \omega t \tag{4.39}$$

The average power is obtained by determining the mean value of the waveform shown, which is given by

$$P = (\omega/2\pi) \int_0^{2\pi/\omega} V_m I_m \sin^2 \omega t \, dt = (\omega/2\pi) V_m I_m \int_0^{2\pi/\omega} \sin^2 \omega t \, dt$$

Using the identity $\sin^2 \theta = (1 - \cos 2\theta)/2$ we have

$$P = (\omega/2\pi) V_m I_m = \int_0^{2\pi/\omega} [(1 - \cos 2\omega t)/2] \, dt = (V_m I_m \omega)/4\pi [t - (\sin 2\omega t/2\omega)]_0^{2\pi/\omega}$$

98 Single-phase a.c. circuits

Note that this indicates a frequency of twice that of the supply voltage, confirming the evidence of the waveforms of Fig. 4.37.

Putting in the limits, this reduces to $V_m I_m \omega/4\pi[2\pi/\omega] = V_m I_m/2$. Rewriting this as $(V_m/\sqrt{2})(I_m/\sqrt{2})$ we convert the maximum values to rms values and we have for the mean power

$$P = VI \tag{4.38 bis}$$

Since $V = IR$ we may also write

$$P = I^2 R \tag{4.40}$$

and

$$P = V^2/R \tag{4.41}$$

Purely inductive circuits

For the purely inductive circuit of Fig. 4.8 the voltage, current and power waveforms are given in Fig. 4.38. Note that the power waveform is sinusoidal and has equal positive and negative half cycles. The average is therefore zero.

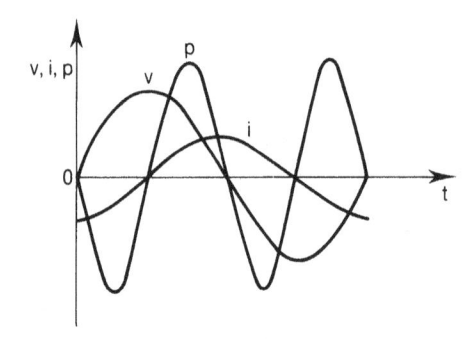

Figure 4.38

If the voltage is represented by $v = V_m \sin \omega t$ then the current will be given by $i = -I_m \cos \omega t$ since it is 90° lagging the voltage. The instantaneous power is then given by

$$p = vi = -V_m I_m \sin \omega t \cos \omega t \tag{4.42}$$

The average power is $P = -(\omega/2\pi) \int_0^{2\pi/\omega} (V_m I_m \cos \omega t \sin \omega t) dt$.

Using the identity $\sin 2\theta = 2 \sin \theta \cos \theta$ we get

$$P = -(\omega/2\pi) V_m I_m \int_0^{2\pi/\omega} [(\sin 2\omega t)/2] dt = -(\omega/4\pi) V_m I_m [-\cos 2\omega/2\omega]_0^{2\pi/\omega} = 0$$

Again the power frequency is twice the supply frequency.

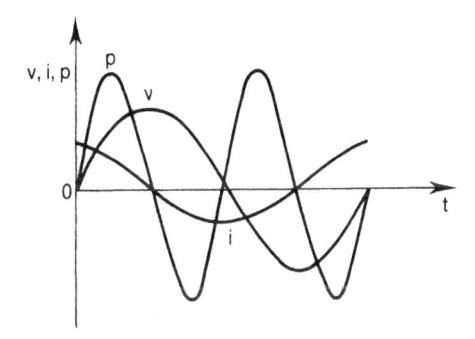

Figure 4.39

Purely capacitive circuits

For the circuit of Fig. 4.11 the voltage, current and power waveforms are given in Fig. 4.39. If the voltage is given by $v = V_m \sin \omega t$ then the current, being 90° ahead of the voltage, will be given by $i = I_m \cos \omega t$ and the instantaneous power is

$$p = V_m I_m \sin \omega t \cos \omega t \tag{4.43}$$

Since Equations (4.42) and (4.43) are identical mathematically then calculation of the mean power will yield the same result as for the purely inductive circuit. The mean power in a purely capacitive circuit is therefore zero.

Resistive–reactive circuits

For circuits which contain resistance together with inductive and/or capacitive reactance there will be a phase angle ϕ which in general lies between 0° and 90°. In these cases, if the voltage is represented by $v = V_m \sin \omega t$ then the current will be given by $i = I_m \sin(\omega t + \phi)$ where ϕ can be positive or negative. The instantaneous power is then $p = vi = V_m \sin \omega t \, I_m \sin(\omega t + \phi)$.

The average power is

$$P = (\omega/2\pi) \int_0^{2\pi/\omega} V_m \sin \omega t \, I_m \sin(\omega t + \phi) dt$$

$$= V_m I_m \omega/2\pi \int_0^{2\pi/\omega} (\sin \omega t [\sin \omega t \cos \phi + \cos \omega t \sin \phi]) dt$$

$$= V_m I_m \omega/2\pi \int_0^{2\pi/\omega} (\sin^2 \omega t \cos \phi + \sin \omega t \cos \omega t \sin \phi]) dt$$

$$= V_m I_m \omega/2\pi \int_0^{2\pi/\omega} (\sin^2 \omega t \cos \phi) dt + V_m I_m \omega/2\pi \int_0^{2\pi/\omega} (\sin \omega t \cos \omega t \sin \phi) dt$$

From the analysis of the purely resistive circuit, we see that the first term of the right-hand side of this equation reduces to $VI \cos \phi$. From the analysis of purely reactive circuits and Equation (4.42) we see that the second integral is zero. The average power is therefore given by $P = VI \cos \phi$.

100 Single-phase a.c. circuits

In general, then, in a single-phase sinusoidal a.c. circuit for which the rms value of the supply voltage is V, the rms value of the supply current is I and the phase angle of the circuit is ϕ, the power is given by

$$P = VI \cos \phi \tag{4.44}$$

For a purely resistive circuit $\phi = 0$ and $\cos \phi = 1$ so that the power is VI, which agrees with the result obtained previously. For a purely reactive circuit $\phi = 90°$ and $\cos \phi = 0$ so that the power is zero which agrees with the results obtained previously.

Power components

Fig. 4.40 shows the phasor diagram for a circuit having a lagging phase angle of ϕ. The current I is shown to have two components at right angles. These are $I \cos \phi$ in phase with V and $I \sin \phi$ lagging V by 90°. If we multiply all three

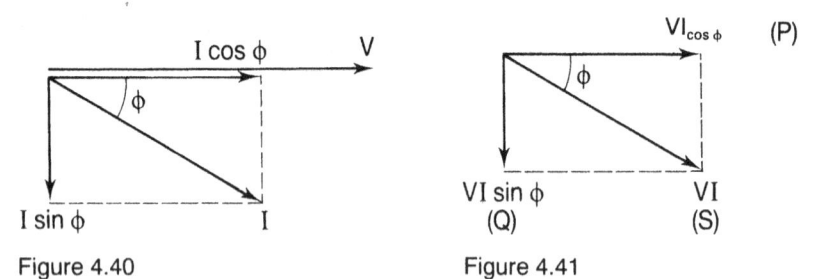

Figure 4.40

Figure 4.41

currents by V we obtain the phasor diagram of Fig. 4.41 and, in this diagram: VI is a phasor representing the so-called apparent power (symbol S) which is measured in volt-amperes (VA); $VI \cos \phi$ is a phasor representing the real power (symbol P) which is measured in watts (W); and $VI \sin \phi$ is a phasor representing the reactive power (symbol Q) which is measured in volt-amperes reactive (Var).

In complex form we have, for a lagging phase angle

$$S = P - jQ \tag{4.45}$$

For a leading phase angle

$$S = P + jQ \tag{4.46}$$

The magnitude of S is obtained from

$$S = \sqrt{(P^2 + Q^2)} \tag{4.47}$$

The phase angle is obtained from

$$\phi = \tan^{-1}(Q/P) \tag{4.48}$$

Also

$$\phi = \sin^{-1}(Q/S) \tag{4.49}$$

and

$$\phi = \cos^{-1}(P/S) \tag{4.50}$$

Since $V = IZ$ and $\cos \phi = R/Z$ then

$$P = VI \cos \phi = (IZ)I(R/Z) = I^2R \quad \text{watts} \tag{4.51}$$

Also, since $\sin \phi = X/Z$ then $Q = VI \sin \phi = (IZ)I(X/Z)$ and

$$Q = I^2X \quad \text{volt-amperes reactive} \tag{4.52}$$

Power factor

The real power (P) in a circuit is obtained by multiplying the apparent power (S) by a factor $\cos \phi$ which is called the power factor of the circuit.

$$P = S \cos \phi \tag{4.53}$$

$$\text{Power factor} = \cos \phi = P/S \tag{4.54}$$

Example 4.22

The circuit of Fig. 4.42 is fed from a 12 V, 50 Hz supply. Calculate (1) the current, (2) the reactive power, (3) the power factor.

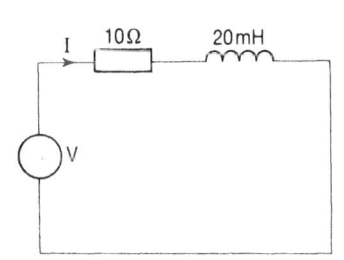

Figure 4.42

Solution

1 The impedance is

$$Z = R + jX_L = 10 + j2\pi fL = 10 + j2\pi\, 50 \times 20 \times 10^{-3} = (10 + j6.28)\ \Omega$$

$$Z = \sqrt{(10^2 + 6.28^2)} = 11.81\ \Omega$$

The current is $I = V/Z = 12/11.81 = 1.02$ A.

2 From Equation (4.52) the reactive power is
  $Q = I^2 X_L = 1.02^2 \times 6.28 = 6.53$ Var

3 The power factor is $\cos \phi = R/Z = 10/11.81 = 0.846$ lagging.

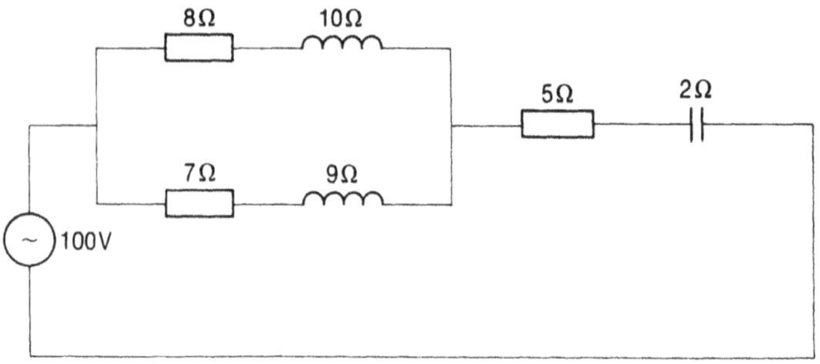

102 Single-phase a.c. circuits

Figure 4.43

Example 4.23

For the series–parallel circuit of Fig. 4.43 determine:

(1) the equivalent impedance of the circuit;

(2) the total power consumed by the circuit;

(3) the reactive power in the capacitive reactance;

(4) the overall power factor of the circuit.

Solution

1 The impedance of the upper inductive branch is

$$Z_1 = (8 + j10) \,\Omega = \sqrt{(8^2 + 10^2)} \angle \tan^{-1}(10/8) \,\Omega$$

so

$$Z_1 = 12.81 \angle 51.34° \,\Omega$$

The impedance of the lower inductive branch is

$$Z_2 = (7 + j9) \,\Omega = \sqrt{(7^2 + 9^2)} \angle \tan^{-1}(9/7) \,\Omega$$

so

$$Z_2 = 11.4 \angle 52.13° \,\Omega$$

Now

$$Z_1 + Z_2 = (8 + 7) + j(10 + 9)$$
$$= (15 + j19) \,\Omega = \sqrt{(15^2 + 19^2)} \angle \tan^{-1}(19/15) \,\Omega$$

so

$$Z_1 + Z_2 = 24.2 \angle 51.7° \,\Omega$$

The equivalent impedance of the parallel inductive branches is

$$Z_1Z_2/(Z_1 + Z_2) = [(12.81 \times 11.4)/24.2]\angle(51.34 + 52.13 - 51.7)$$
$$= 6.03\angle 51.77° \; \Omega$$
$$= 6.03(\cos 51.77 + j \sin 51.77) = (3.73 + j4.73) \; \Omega$$

The impedance of the capacitive branch is

$$Z_3 = (5 - j2) \; \Omega$$

The impedance of the complete circuit,

$$Z_{eq} = Z_3 + Z_1Z_2/(Z_1 + Z_2) = (5 + 3.73) + j(-2 + 4.73) \; \Omega$$

so

$$Z_{eq} = (8.73 + j2.73) = \sqrt{(8.73^2 + 2.73^2)}\angle \tan^{-1}(2.73/8.73) = 9.15\angle 17.36° \; \Omega$$

2 The current drawn from the supply is

$$V/Z_{eq} = (100/9.15)\angle 0 - 17.36 = 10.92\angle -17.36° \; A$$

The total power consumed is

$$P = VI \cos \phi = 100 \times 10.92 \cos 17.36 = 1042 \; W$$

3 The reactive power in the capacitive reactance is

$$Q = I^2 X_C = 10.92^2 \times 2 = 238 \; Var$$

4 The equivalent reactance of the whole circuit is positive, indicating an effective inductive reactance so that the overall power factor is lagging and its value is given by $\cos \phi = \cos 17.36 = 0.9544$.

4.8 SELF-ASSESSMENT TEST

1 Define an alternating quantity.

2 Give the unit of frequency.

3 Write down the relationship between frequency (f) and periodic time (T).

4 What is the peak value of the alternating quantity represented by $25 \sin \omega t$?

5 What is the frequency of the alternating quantity represented by $50 \cos 314 \, t$?

6 If $v = V \sin 2\pi f t$ and $i = I \cos 2\pi f t$ state the phase difference between v and i.

7 What is the rms value of a sinusoidal voltage whose maximum value is 141 V?

104 *Single-phase a.c. circuits*

8 Determine the form factor of an alternating waveform whose rms value is 50 and whose average value is 45.

9 Give the relationship between the voltage and current in a purely resistive a.c. circuit.

10 Calculate the inductive reactance of a coil whose inductance is 60 mH when it is connected to a supply of frequency 400 Hz.

11 A capacitor of capacitance 0.1 μF is connected to a 50 Hz supply. Calculate its capacitive reactance.

12 Give the unit of impedance.

13 Determine the phase angle of a coil having a resistance equal to half its inductive reactance.

14 Determine the impedance of a coil having a resistance and reactance both equal to 20 Ω.

15 Calculate the impedance of a circuit consisting of a capacitor in series with the coil of Question 14 if the capacitive reactance of the capacitor is (a) 10 Ω and (b) 20 Ω.

16 Describe an Argand diagram.

17 Express j^7 in its simplest form.

18 In which quadrant does $(-6 - j5)$ lie?

19 Determine the magnitude and phase angle of the phasor represented by $v = (10 - j15)$.

20 Express $25\angle 30°$ in rectangular coordinate form.

21 Express $3 + j4$ in polar coordinate form.

22 An a.c. circuit consists of a resistor of 5 Ω resistance in series with an inductive reactance of 6 Ω. Express the impedance of this circuit in j form.

23 A resistance of 10 Ω is connected in parallel with a capacitive reactance of 25 Ω. Express the admittance of this circuit in j form.

24 Give an expression using the j-notation which relates real power, reactive power and apparent power.

25 An RLC series circuit has $R = 5$ Ω, $X_L = 6$ Ω and $X_C = 10$ Ω. A current given by $(5 + j0)$ A flows through the circuit. Determine: (a) the apparent power; (b) the real power; (c) the reactive power; (d) the power factor; (e) the applied voltage in j-form.

4.9 PROBLEMS

1. A circuit is supplied from 50 Hz mains whose voltage has a maximum value of 250 V and takes a current whose maximum value is 5 A. At a particular instant ($t = 0$) the voltage has a value of 200 V and the current is then 2 A. Obtain expressions for the instantaneous values of voltage and current as functions of time and determine their values at an instant $t = 0.015$ s. Determine also the phase difference between them.

2. Three circuit elements are connected in series and the voltages across them are given by $v_1 = 50 \sin \omega t$; $v_2 = 40 \sin (\omega t + 60°)$; $v_3 = 60 \sin (\omega t - 30°)$. Determine the total voltage across the series combination and its phase angle with respect to v_1.

3. A circuit consists of an inductance of 0.318 H and a resistance of 5 Ω connected in series. Calculate the resistance, reactance and impedance when it operates on a supply of frequency (a) 25 Hz, (b) 50 Hz.

4. A coil of resistance 10 Ω and inductance 47.7 mH is connected to a 200 V, 50 Hz supply. Calculate (a) the current drawn from the supply, (b) the power factor of the circuit.

5. A capacitor of 50 μF capacitance is connected in series with a resistance of 50 Ω to a 200 V, 50 Hz supply. Calculate (a) the current drawn from the supply, (b) the voltage across the resistance, and (c) the voltage across the capacitance.

6. A capacitor of 80 μF capacitance takes a current of 1 A when supplied with 250 V (rms). Determine (a) the frequency of the supply, and (b) the resistance which must be connected in series with this capacitor in order to reduce the current to 0.5 A at this frequency.

7. A resistance of 50 Ω is connected in series with a variable capacitor across a 200 V, 50 Hz supply.
   (a) When the capacitance is set to 50 μF calculate (i) the current drawn from the supply, (ii) the voltage across the two elements and (iii) the power factor.
   (b) Find the value of the capacitance when the current is 2 A.
   (c) Determine the value of the capacitance required to give a power factor of 0.866 leading.

8. A circuit consists of a 100 Ω resistance in parallel with a 25 μF capacitor connected to a 200 V, 50 Hz supply. Calculate (a) the current flowing in each branch, (b) the total current drawn from the supply, (c) the impedance of the circuit, and (d) the phase angle of the circuit.

9. A capacitor having a reactance of 5 Ω is connected in series with a resistor of 10 Ω. This circuit is then connected (a) in series and (b) in parallel with

a coil of impedance $(5 + j7)$ Ω. Calculate for each case (i) the current drawn from the supply, (ii) the power supplied, and (iii) the power factor of the whole circuit.

10 A coil of inductance 15.9 mH and resistance 9 Ω is connected in parallel with a coil of inductance 38.2 mH and resistance 6 Ω across a 200 V, 50 Hz supply. Determine (a) the conductance, susceptance and admittance of the circuit, (b) the current drawn from the supply and (c) the total power consumed in kW.

11 Three coils are connected in parallel across a 200 V supply. Their impedances are $(10 + j30)$ Ω, $(20 + j0)$ Ω and $(1 - j20)$ Ω. Determine (a) the current drawn from the supply and (b) the power factor of the circuit.

12 Determine (a) the current drawn from the supply and (b) the total power consumed in the circuit of Fig. 4.44.

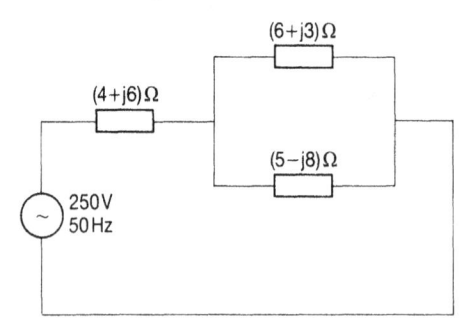

Figure 4.44

5 Three-phase a.c. circuits

5.1 INTRODUCTION

Three-phase has a number of advantages over single-phase:

- A three-phase machine of a given physical size gives more output than a single-phase machine of the same size and most electrical power generation is carried out by means of three-phase synchronous generators.

- There is a considerable amount of saving in conductor material to be gained by using three-phase rather than single-phase for the purposes of power transmission by overhead lines or underground cables.

- The three-phase induction motor is the cheapest and most robust of machines and accounts for the vast majority of the world's industrial machines.

5.2 GENERATION OF THREE-PHASE VOLTAGES

We saw in Chapter 4 that a single-phase a.c. voltage is generated by rotating a single coil in a magnetic field. A three-phase a.c. system is generated by rotating three coils in a magnetic field, the coils being mutually displaced in space by $2\pi/3$ radians (120°) as shown in Fig. 5.1(a). The waveforms of the three-phase system of voltages thus produced are shown in Fig. 5.1(b). The coils in which the voltages are generated constitute the armature, while the system producing the magnetic field is called the field system. In the large generators found in power stations the armature system is stationary and it is the field system which is made to rotate.

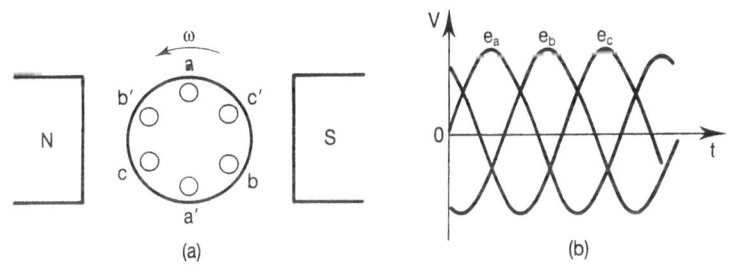

Figure 5.1

Coil A has two ends labelled a and a'; coil B has two ends b and b'; coil C has ends c and c'. End b of coil B is displaced by $2\pi/3$ radians from end a of coil A, and end c of coil C is displaced by $2\pi/3$ radians from end b of coil B (in a clockwise direction). This means that if the coils are rotated in an anticlockwise direction at an angular frequency of ω radians per second then coil A passes the N-pole of the magnetic field $2\pi/3\omega$ seconds ahead of coil B which in turn passes the N-pole $2\pi/3\omega$ seconds ahead of coil C. The phasor diagram of the voltages is given in Fig. 5.2 and shows the symmetry of the system.

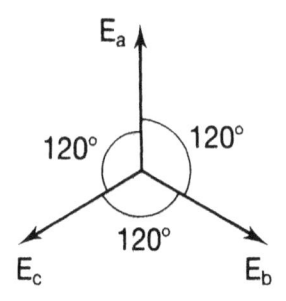

Figure 5.2

5.3 PHASE SEQUENCE

The order in which the coils pass a given point in an anticlockwise direction is called the positive phase sequence of the three-phase system, and in the case shown, in which coil A generates phase A, coil B generates phase B and coil C generates phase C, the phase sequence (the word positive is understood) is ABC. The negative phase sequence of this system is ACB. The phase sequence of any system can be reversed by reversing the connections to two of the coils (say B and C) as shown in Fig. 5.3(a) and (b).

Reversing the connections to any one coil simply upsets the symmetrical nature of the system as illustrated in Fig. 5.4(a) and (b).

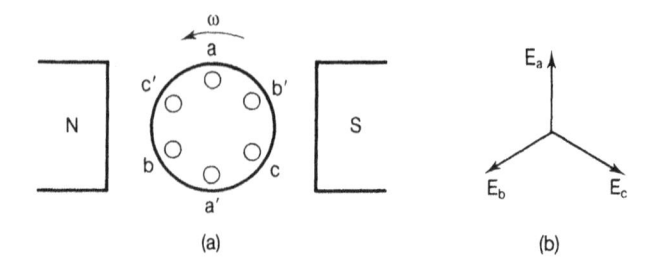

Figure 5.3

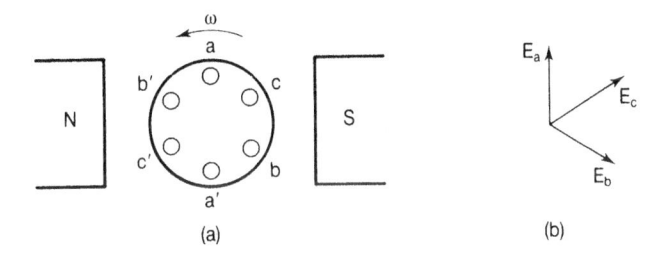

Figure 5.4

5.4 BALANCED THREE-PHASE SYSTEMS

It is important that the system is not only symmetrical (i.e. the voltage phasors are mutually displaced by 120°) but also balanced (i.e. the voltages are equal in magnitude so that the phasors are of equal length). This can be achieved by ensuring that the coils are identical and that their mutual 120° separation is maintained. If this is done then the three generated voltages may be represented as follows:

- $E_A = E_m \sin \omega t$ where E_m is the maximum value of the generated emf;
- $E_B = E_m \sin (\omega t - 2\pi/3)$ since the emf generated in coil B lags that in coil A by 120°;
- $E_C = E_m \sin (\omega t - 4\pi/3)$ since the emf generated in coil C lags that in coil A by 240°.

We could also express E_C as $E_m \sin (\omega t + 2\pi/3)$ because E_C is also 120° ahead of E_A. Phasorially this is represented as shown in Fig. 5.5 and the phasors can be drawn to represent E_m or E (the rms values).

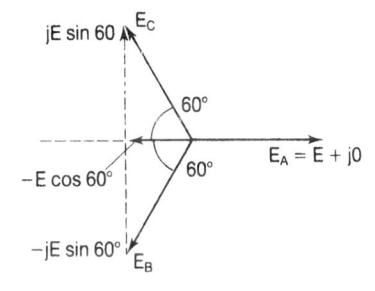

Figure 5.5

Note also that with E_A as the reference phasor, we can use the complex (j) notation to write

- $E_A = E(1 + j0)$

- $E_B = -E \cos 60° - jE \sin 60° = E(-0.5 - j0.866)$
- $E_C = -E \cos 60° + jE \sin 60° = E(-0.5 + j0.866)$.

A balanced system is one in which the load impedance on each phase is the same in quality and quantity so that if the load impedances are $Z_A(=R_A + jX_A)$, $Z_B(=R_B + jX_B)$ and $Z_C(=R_C + jX_C)$ then $R_A = R_B = R_C$ ($=R$ say) and $X_A = X_B = X_C$ ($=X$ say). The phase angles ϕ_A ($=\tan^{-1}(X_A/R_A)$), $\phi_B(=\tan^{-1}(X_B/R_B))$ and $\phi_C(=\tan^{-1}(X_C/R_C))$ are also equal (to ϕ say).

Six-wire system

If a load is connected to each of the phases separately, six wires would be needed as shown in Fig. 5.6.

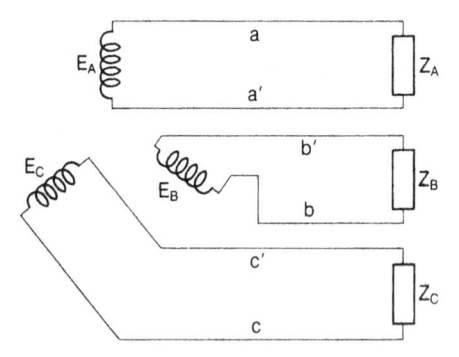

Figure 5.6

Four-wire system

By connecting together the three return wires (a', b' and c') of the six-wire system they could be replaced by just one wire making a total of four wires. The common return wire is called the neutral wire and the junction of the three coils is called the neutral point. This connection, shown in Fig. 5.7, is called a four-wire star connection.

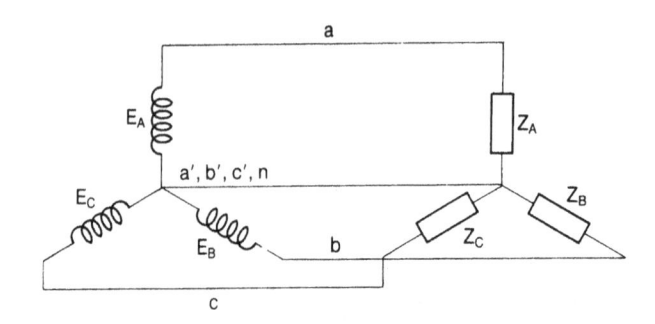

Figure 5.7

Three-wire star system

If the system is balanced the current in each phase can be represented by three sine waveforms, each having the same maximum value and with a successive phase displacement of 120°. At any instant therefore the sum of the currents is zero. Because the current in the return wire at any instant $(i_A + i_B + i_C) = 0$ it carries no current and may be dispensed with entirely, giving the three-wire system shown in Fig. 5.8.

The current flowing in the generator-phase winding or in the load-phase impedance is called the phase current (I_{PH}). The current flowing in the wires connecting the generator to the load is called the line current (I_L). Clearly in this connection, the phase current and the line current are one and the same so

$$I_{PH} = I_L \tag{5.1}$$

The junction of the three-phase windings (or of the three-phase loads) is called the star point. The voltage between any line and the star point is called the phase voltage (E_{PH} on the generator side; V_{PH} on the load side). The voltage between any two lines is called the line voltage (E_L on the generator side; V_L on the load side). In order to obtain the relationship between the phase voltage and the line voltage we note from Fig. 5.8 that the line voltage is $V_{AB} = V_{AN} - V_{BN}$. This is shown phasorially in Fig. 5.9 where we have drawn $-V_{BN}$ equal and opposite to V_{BN} and added it to V_{AN} to give V_{AB}. A line is drawn perpendicularly from the end of V_{AN} to meet V_{AB} at M, which, by the geometry of the diagram, is its mid-point. Note also that the angle θ is 30°.

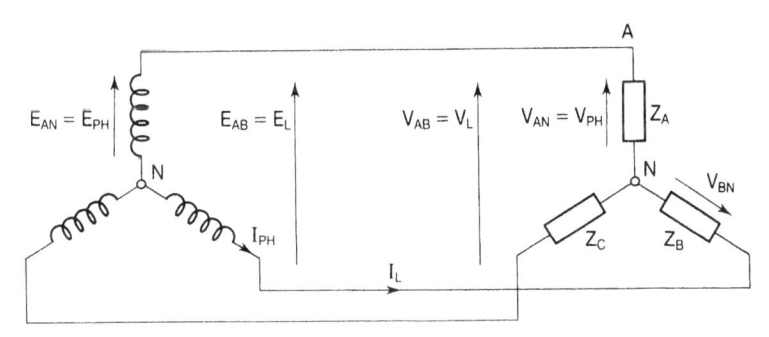

Figure 5.8

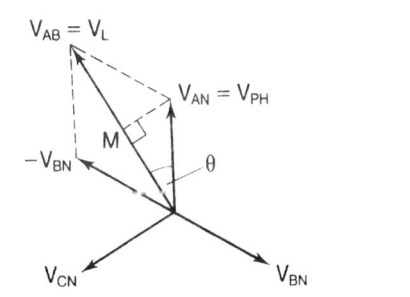

Figure 5.9

112 Three-phase a.c. circuits

Now

$$V_{AB} = 2 V_{AN} \cos \theta = 2 V_{AN} \cos 30° = 2 V_{AN} \sqrt{3}/2 = \sqrt{3} V_{AN}$$

But $V_{AB} = V_L$ and $V_{AN} = V_{PH}$ so that

$$V_L = \sqrt{3} V_{PH} \tag{5.2}$$

Summarizing, we can say that in a balanced three-phase star connection:

- the line current equals the phase current ($I_L = I_{PH}$);
- the line voltage is $\sqrt{3}$ times the phase voltage and leads it by 30° ($V_L = \sqrt{3} V_{PH} \angle 30°$);
- the phase current is equal to the phase voltage divided by the phase impedance ($I_{PH} = V_{PH}/Z_{PH}$);
- the phase angle is given by $\phi = \tan^{-1} (X_L/R_L)$ where X_L and R_L are the load reactance and resistance, respectively;
- the power factor is given by $\cos \phi$.

It is also worth noting here that in a star connection the star point is available which means that 'mixed' (three-phase and single-phase) loads can be supplied from the same supply. This is particularly useful in power distribution networks.

Example 5.1

Three 10 Ω resistors are connected in star to a three-phase supply whose line voltage is 440 V. Calculate (1) the voltage across each resistor, (2) the line current, (3) the current in each phase if one of the resistors becomes open circuited.

Solution

The connection is shown in Fig. 5.10.

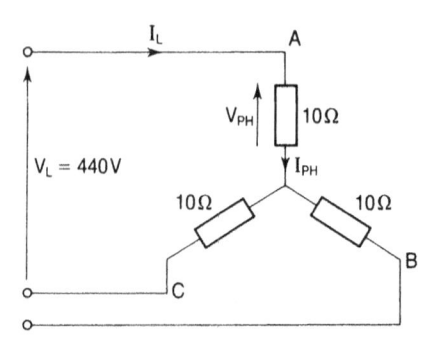

Figure 5.10

1 The voltage across each resistor is the phase voltage, and from Equation (5.1)

$$V_{PH} = V_L/\sqrt{3} = 440/\sqrt{3} = 254 \text{ V}$$

2 The phase current is

$$I_{PH} = V_{PH}/Z_{PH} = 254/10 = 25.4 \text{ A}$$

Since this is a star connection the line current is $I_L = I_{PH} = 25.4$ A.

3 Let the resistor in the B phase become open circuited. The circuit now consists of the other two resistors connected in series across 440 V. The current in line B is $I_L = 0$. The current in line A equals that in line C:

$$I_B = I_C = 440/(10 + 10) = 22 \text{ A}$$

Example 5.2

Three identical coils having a resistance of 3 Ω and an inductive reactance of 4 Ω are connected in star to a three-phase supply whose line voltage is 240 V. Determine (1) the voltage across each coil, (2) the current drawn from the supply, (3) the power factor.

Solution

Fig. 5.11 shows the circuit.

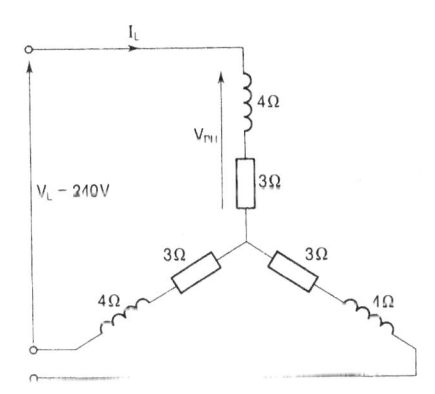

Figure 5.11

1 $V_{PH} = V_L/\sqrt{3} = 240/\sqrt{3} = 139$ V.

2 $I_L = I_{PH} = V_{PH}/Z_{PH} = 139/\sqrt{(3^2 + 4^2)} = 139/5 = 27.8$ A.

3 The power factor is $\cos \phi = R/Z = 3/5 = 0.6$ lagging.

Delta system

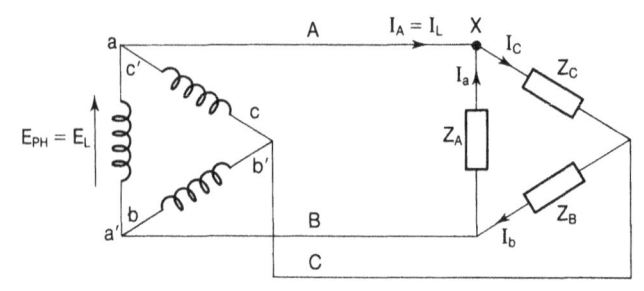

Figure 5.12

If, instead of connecting together the corresponding ends of the three coils (say all the 'start' ends a, b and c, or all the 'finish' ends a', b' and c') the finish end of one coil were connected to the start end of the next in order (a' to b; b' to c; c' to a) we obtain the so-called delta connection shown in Fig. 5.12. In this case it is clear that the line voltages are the same as the phase voltages because each phase is connected directly between two lines:

$$V_{PH} = V_L \tag{5.3}$$

Although the phase voltages act around the delta they sum to zero at any instant since

$$e_{AB} + e_{BC} + e_{CA} = E \sin \omega t + E \sin (\omega t - 2\pi/3) + E \sin (\omega t - 4\pi/3) = 0$$

There is therefore no circulating current around the delta.

The line currents are different from the phase currents and applying KCL to node X, for example, we see that the line current I_A is the difference of the two phase currents I_c and I_a. This is shown in the phasor diagram of Fig. 5.13. In this diagram we have added $-I_a$ to I_c to get I_A. Now

$$I_A = 2I_{AB} \cos \theta = 2I_{AB} \cos 30°$$

(the geometry being the same as that in Fig. 5.9), so

$$I_A = 2I_{AB}\sqrt{3}/2 = \sqrt{3}I_{AB}$$

But $I_A = I_L$ and $I_c = I_{PH}$ so that

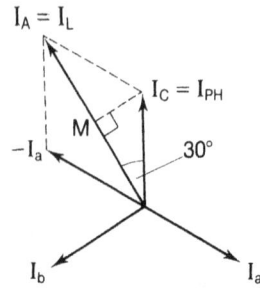

$$I_L = \sqrt{3}I_{PH} \tag{5.4}$$

Figure 5.13

Also there is a phase difference of 30° between the line and phase currents.

Example 5.3

Three impedances, each of impedance $Z = (5 - j12)\ \Omega$ are connected in delta across a three-phase supply whose line voltage is 110 V. Determine (1) the voltage across each impedance, (2) the line current drawn from the supply, (3) the power factor.

Solution

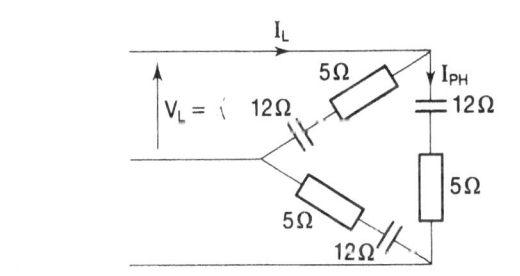

Figure 5.14

The arrangement is shown in Fig. 5.14.

1. The voltage across each impedance is the phase voltage which, for a delta connection, is the same as the line voltage: $V_{PH} = V_L = 110$ V.

2. The phase current is

$$I_{PH} = V_{PH}/Z_{PH} = 110/\sqrt{(5^2 + 12^2)} = 110/13 = 8.46\ A$$

3. The power factor is $\cos \phi = R/Z = 5/13 = 0.385$ leading.

5.5 POWER IN BALANCED THREE-PHASE CIRCUITS

The total power (P) in any three-phase system whose phases are A, B and C is given by the sum of the powers in each of the three phases. If these are respectively P_A, P_B, and P_C then

$$P = P_A + P_B + P_C \tag{5.5}$$

For a balanced system the power in each phase is the same, so that the total power is simply three times the power in one phase.

We saw in Chapter 4 that the power in a single-phase circuit is given by $P = VI \cos \phi$ where V and I are the rms voltage and current, respectively, and $\cos \phi$ is the power factor. In a balanced three-phase circuit therefore the total power is given by:

$$P = 3 V_{PH} I_{PH} \cos \phi \tag{5.6}$$

where V_{PH} is the rms value of the phase voltage, I_{PH} is the rms value of the phase current, and ϕ is the angle between V_{PH} and I_{PH}.

For the balanced star-connected system we have that $V_{PH} = V_L/\sqrt{3}$ and that $I_{PH} = I_L$ so that $P = 3(V_L/\sqrt{3})I_L \cos \phi = \sqrt{3} V_L I_L \cos \phi$.

For a balanced delta-connected system we have that $V_{PH} = V_L$ and that $I_{PH} = I_L/\sqrt{3}$ so that $P = 3 V_L(I_L/\sqrt{3}) \cos \phi = \sqrt{3} V_L I_L \cos \phi$.

For any balanced three-phase system, therefore, whether it be star or delta connected, the total power is given by

$$P = \sqrt{3} V_L I_L \cos \phi \quad \text{watts} \tag{5.7}$$

In this equation, V_L and I_L are the rms values of the line voltage and current, respectively, and $\cos \phi$ is the power factor of the circuit.

From Equation (4.53), Chapter 4, the real power (P) is the apparent power (S) multiplied by the power factor. It follows that $S = P/\cos \phi$, so the apparent power is then given by

$$S = \sqrt{3} V_L I_L \quad \text{volt-amperes} \tag{5.8}$$

We also saw in Chapter 4 that the reactive power (Q) is the apparent power (S) multiplied by the sine of the phase angle ϕ so that

$$Q = \sqrt{3} V_L I_L \sin \phi \quad \text{volt-amperes reactive} \tag{5.9}$$

Example 5.4

A three-phase load takes a line current of 10 A at 0.8 power factor lagging from a supply whose line voltage is 415 V. Determine (1) the power taken by the load, (2) the weekly energy cost of operating the load for eight hours a day, six days a week. The energy charge is 7.5 p per unit.

Solution

1 From Equation (5.7) the power taken is given by $P = \sqrt{3}V_L I_L \cos \phi$. In this case $V_L = 415$ V, $I_L = 10$ A and $\cos \phi$ (the power factor) is 0.8. Therefore

 $P = \sqrt{3} \times 415 \times 10 \times 0.8 = 5750$ W

2 The energy consumed by the load is measured in joules, which is watt-seconds. This is a rather small unit, so that for most purposes supply authorities use the kilowatt-hour (kW-hr) as the unit of energy. Energy used per week is

 $P \times (6 \times 8) = 5750 \times 48 = 276011$ W-hr $= 276.011$ kW-hr

 Cost of energy is $276.011 \times 7.5 = £20.70$. This is the cost of the electricity consumed. In operating any piece of equipment there are also costs of maintenance and depreciation.

Measurement of power in balanced three-phase circuits

For a four-wire system it is only necessary to use just one wattmeter connected as shown in Fig. 5.15. A wattmeter is calibrated to read the product $VI \cos \phi$ where V is the voltage across its voltage coil, I is the current through its current coil and ϕ is the angle between them. The wattmeter in Fig. 5.15 has the phase voltage across its voltage coil and the phase current (which is also the line current in this case) through its current coil. It will therefore read the power in one phase and so the total power is obtained by multiplying the reading by three.

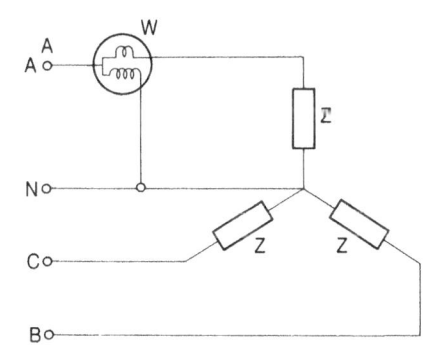

Figure 5.15

For a three-wire system, however, the so-called two-wattmeter method is used, the two wattmeters being connected as shown in Fig. 5.16. Since a wattmeter reads the product of the voltage across its voltage coil with the current through its current coil and the cosine of the angle between them, then

- W_1 will read $V_{AC}I_A \cos \phi_1$ (ϕ_1 is the angle between V_{AC} and I_A)
- W_2 will read $V_{BC}I_B \cos \phi_2$ (ϕ_2 is the angle between V_{BC} and I_B).

The phasor diagram is drawn in Fig. 5.17 assuming a lagging power factor of

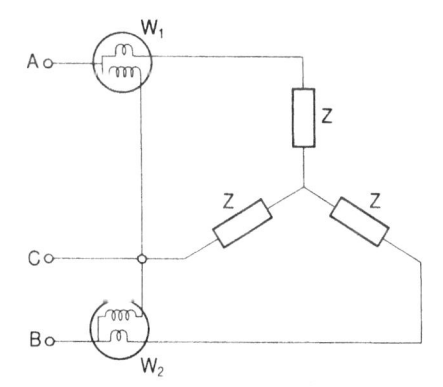

Figure 5.16

118 Three-phase a.c. circuits

$\cos \phi$ so that the phase currents are lagging the phase voltages by ϕ. Wattmeter W_1 reads

$$V_{AC}I_A \cos (30° - \phi) \qquad (5.10)$$

where V_{AC} is a line voltage and I_A is a line current.

Wattmeter W_2 reads

$$V_{BC}I_B \cos (30° + \phi) \qquad (5.11)$$

where V_{BC} is a line voltage and I_B is a line current.

For a leading power factor of $\cos \phi$ the sign of ϕ will change in Equations (5.10) and (5.11). The power represented by the reading on W_1 is:

$$P_1 = V_L I_L (\cos 30° \cos \phi + \sin 30° \sin \phi) = V_L I_L [(\sqrt{3}/2) \cos \phi + (1/2)\sin \phi]$$

The power represented by the reading on W_2 is:

$$P_1 = V_L I_L (\cos 30° \cos \phi - \sin 30° \sin \phi) = V_L I_L [(\sqrt{3}/2) \cos \phi - (1/2)\sin \phi]$$

The power represented by the readings on W_1 and W_2 is thus

$$P_1 + P_2 = \sqrt{3}\, V_L I_L \cos \phi \qquad (5.12)$$

which is the total power in a balanced three-phase circuit.

Therefore the sum of the readings on the two wattmeters gives the total power in the three-phase circuit. If the phase angle is greater than 60° (leading or lagging) one of the wattmeters will read negative because $\cos (30° + \phi)$ is then negative and the reading must be subtracted from the other to give the total power.

Now $P_1 - P_2 = V_L I_L \sin \phi$, which is $1/\sqrt{3}$ of the total reactive power, so if we multiply $(P_1 - P_2)$ by $\sqrt{3}$ we obtain the total reactive power in the three-phase circuit. Thus

$$\text{Var} = \sqrt{3}(P_1 - P_2) = \sqrt{3}\, V_L I_L \sin \phi \qquad (5.13)$$

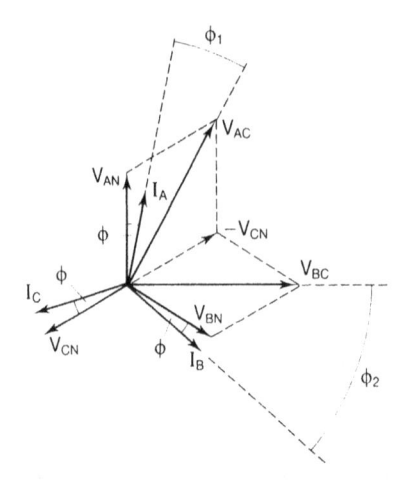

Figure 5.17

Since the phase angle ϕ is given by $\tan^{-1}$ (reactive power/real power) then

$$\phi = \tan^{-1}[\sqrt{3}(P_1 - P_2)/(P_1 + P_2)] \qquad (5.14)$$

The power factor is then simply $\cos \phi$.

Summarizing, using the two-wattmeter method in any balanced star or delta connected three-phase circuit in which the readings are P_1 and P_2, we can obtain the following information:

- total real power

$$W = (P_1 + P_2) \text{ watts} \qquad (5.12 \text{ bis})$$

- total reactive power

$$Q = \sqrt{3}(P_1 - P_2) \text{ Var} \qquad (5.13 \text{ bis})$$

- power factor

$$\cos \phi = \cos\{\tan^{-1}[\sqrt{3}(P_1 - P_2)/(P_1 + P_2)]\} \qquad (5.15)$$

Example 5.5

Two wattmeters are connected to measure the power input to a 400 V, 50 Hz, three-phase motor running on full load with an efficiency of 90 per cent. The readings on the two wattmeters are 30 kW and 10 kW. Calculate (1) the input power to the motor, (2) the reactive power, (3) the power factor, (4) the useful output power from the motor.

Solution

1 Let the readings on the two wattmeters be P_1 and P_2. Then, from Equation (5.12), the power to the motor is

$$P_1 + P_2 = 30 + 10 = 40 \text{ kW}$$

2 From Equation (5.13) the reactive power is

$$\sqrt{3}(P_1 - P_2) = \sqrt{3} \times 20 = 34.64 \text{ Var}$$

3 From Equation (5.14)

$$\phi = \tan^{-1}[\sqrt{3}(P_1 - P_2)/(P_1 + P_2)] = \tan^{-1}(34.64/40) = 40.89°$$

so the power factor is

$$\cos \phi = \cos 40.89° = 0.756 \text{ lagging}$$

4 The efficiency (η) of the motor is defined as the useful output power (P_o) divided by the total input power (P_i). Thus

$$P_o = \eta P_i = 0.9 \times 40 = 36 \text{ kW}$$

Example 5.6

A 440 V three-phase motor has a useful output of 50 kW and operates at a power factor of 0.85 lagging with an efficiency of 89 per cent. Calculate the readings on two wattmeters connected to measure the input power.

Solution

Since efficiency is $\eta = P_o/P_i$, the input power $P_i = P_o/\eta = 50/0.89 = 56.2$ kW. Let the readings on the two wattmeters be P_1 and P_2. Then $P_1 + P_2 = 56.2$ kW. Now $\phi = \cos^{-1} 0.85 = 31.79°$, so

$\tan \phi = 0.6197$

From Equation (5.14) $\tan \phi = \sqrt{3}(P_1 - P_2)/(P_1 + P_2)$, so

$P_1 - P_2 = \tan \phi \times (P_1 + P_2)/\sqrt{3} = (0.6197 \times 56.2)/\sqrt{3} = 20.1$ kW

We now have

$P_1 + P_2 = 56.2$ kW
$P_1 - P_2 = 20.1$ kW

Adding gives

$2P_1 = 76.3$ kW
$P_1 = 38.15$ kW

It follows that

$P_2 = 18.05$ kW

5.6 SELF-ASSESSMENT TEST

1 Give an advantage of three-phase systems over single-phase systems for the purpose of power transmission.

2 Give an advantage of three-phase systems over single-phase systems for the purpose of power generation.

3 Explain how three-phase emfs are generated.

4 Explain the difference between a *star-connected* system and a *delta-connected* system.

5 Explain why a star connection is more usual for power distribution purposes.

6 State the meaning of the term 'phase sequence' when applied to three-phase systems.

7 State the phase sequence of a system for which $E_a = 240 \sin \omega t$; $E_b = 240 (\sin \omega t + 120°)$; $E_c = 240 \sin (\omega t - 120°)$.

8 How may the phase sequence of a three-phase system be reversed?

9 If the emf generated in phase A of a three-phase system of phase sequence ABC is given by $e = E_M \sin \omega t$ volts, give an expression for the emf in phase C.

10 Explain what is meant by a *balanced* three-phase system.

11 The phase current of a star-connected three-phase system is 25 A. What is the line current?

12 State the phase displacement between the phase voltage and the line voltage of a star connected three-phase system.

13 Give the line current of a delta-connected three-phase system for which the phase current is 10 A.

14 A three-phase delta-connected load has a line voltage of 415 V and a line current of 22 A. The power factor of the load is 0.8 lagging. Determine the power in the load.

15 Give an expression for the apparent power in a three-phase load.

16 The apparent power in a three-phase load is 35 kVA and its power factor is 0.6 lagging. Determine the reactive power.

17 Two wattmeters are connected to measure the total power in a three-phase load and the readings are 240 W and 100 W. What is the total power in the load?

18 Two wattmeters are connected as follows to measure the total power in a three-phase system: one has its current coil connected in the A line and its voltage coil connected between the A and C lines; the other has its current coil connected in the B line and its voltage coil connected between the B and C lines. Give an expression for the reading on each meter in terms of appropriate line voltages, line currents and phase angles.

19 Give an expression for the power factor of a three-phase load in terms of the readings on two wattmeters connected to read the total power.

20 The readings on two wattmeters connected to read the total power in a three-phase system are 125 W and 65 W. Determine (a) the real power, (b) the reactive power, (c) the apparent power, (d) the power factor.

5.7 PROBLEMS

1. A balanced delta-connected load is supplied from a symmetrical three-phase, 400 V source and takes a phase current of 20 A. Determine (a) the line current and (b) the total power consumed.

2. Three identical coils are connected in star and take a total of 3 kW from a three-phase, 200 V supply at a lagging power factor of 0.8. If the same coils are connected in delta and supplied from the same source determine (a) the line current taken and (b) the total power consumed.

3. A balanced star-connected load consisting of a resistance of 1 Ω in series with an inductance of 15 mH in each phase is supplied from a 230 V, 50 Hz supply. Determine the total power consumed in complex form.

4. Two wattmeters connected to measure the power in a three-phase circuit for which the line voltage is 400 V read 40 kW and −10 kW. Determine (a) the power consumed, (b) the power factor and (c) the line current.

5. The power in a three-phase circuit is measured using the two-wattmeter method. Both wattmeters read positively, one reading being twice as big as the other. Calculate the power factor of the circuit.

6. A three-phase motor operates at full load with an efficiency of 90 per cent when supplied from a 2 kV, 50 Hz source. Two wattmeters connected to read the power taken from the supply read 300 kW and 100 kW. Calculate (a) the power input to the motor, (b) the power factor, (c) the line current, and (d) the useful power output from the motor.

6 Resonance

6.1 SERIES RESONANCE

We saw in Chapter 4 that under certain conditions a series RLC circuit such as that shown in Fig. 6.1 behaves as a pure resistance. This happens when the inductive reactance $X_L \, (=2\pi f L)$ is equal to the capacitive reactance $X_C \, (=1/2\pi f C)$. The circuit impedance, $Z \, (=\sqrt{[R^2 + (X_L - X_C)^2]})$ then becomes equal to R since $X_L - X_C = 0$. This condition is called series resonance and has important applications in filter circuits and radio and television tuning circuits.

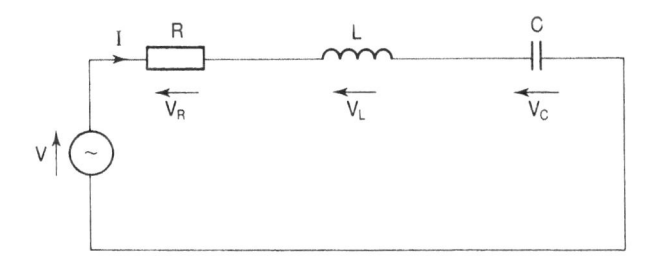

Figure 6.1

The phasor diagram, drawn for the condition when $X_L = X_C$, is given in Fig. 6.2 and this is drawn using the current I as the reference phasor.

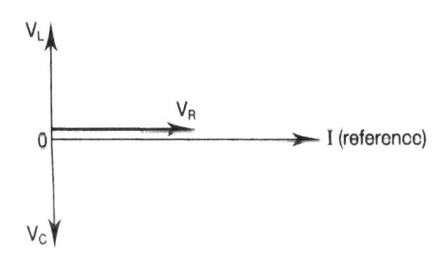

Figure 6.2

- The voltage phasor $V_L = IX_L$ is drawn 90° ahead of the current phasor.
- The voltage phasor $V_C = IX_C$ is drawn 90° behind the current phasor.

124 Resonance

- The voltage phasor $V_R = IR$ is drawn in phase with the current phasor.

The voltage V is the phasor sum of V_L, V_C and V_R and because $V_L = -V_C$ then $V = V_R$. The phase angle of the circuit (the angle between the applied voltage, V, and the current, I) is $\phi = 0$, so that the power factor is $\cos\phi = 1$. The condition $X_L = X_C$ occurs when

$$2\pi fL = 1/2\pi fC \tag{6.1}$$

and this can be made to happen by the variation of L or C or f.

The graphs of X_L and X_C to a base of frequency are shown in Fig. 6.3 and it can be seen that:

(1) as $f \to 0$, $X_L \to 0$ and $X_C \to \infty$

(2) as $f \to \infty$, $X_L \to \infty$ and $X_C \to 0$

(3) at a particular frequency (f_0) $X_L = X_C$ in magnitude.

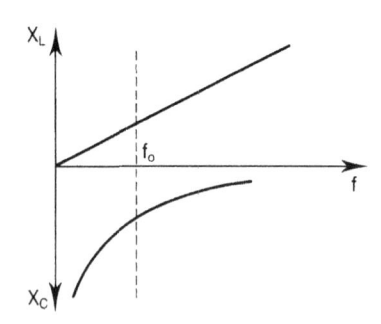

Figure 6.3

This frequency is called the resonant frequency of the circuit and may be calculated from Equation (6.1), putting $f = f_0$. Thus

$$2\pi f_0 L = 1/2\pi f_0 C$$
$$(2\pi f_0)^2 = 1/LC \quad \text{(multiplying both sides by } 2\pi f_0/L\text{)}$$
$$2\pi f_0 = 1/\sqrt{(LC)} \quad \text{(taking the square root of both sides)}$$

Finally

$$f_0 = 1/2\pi\sqrt{(LC)} \tag{6.2}$$

Also, remembering that the angular frequency $\omega = 2\pi f$ we may write

$$\omega_0 = 1/\sqrt{(LC)} \tag{6.3}$$

Summarizing, when an RLC circuit is in a state of series resonance:

- the circuit behaves as a pure resistance;
- the inductive reactance is equal to the capacitive reactance;
- the applied voltage $V = IR$;

- the phase angle is $\phi = 0$ and the power factor is $\cos \phi = 1$;
- the frequency is $f_0 = 1/2\pi\sqrt{(LC)}$ and the angular frequency is $\omega_0 = 1/\sqrt{(LC)}$.

Example 6.1

Calculate the resonant frequency of a circuit consisting of a coil of inductance 50 mH in series with a capacitor of capacitance 200 nF and a resistor of resistance 5 Ω.

Solution

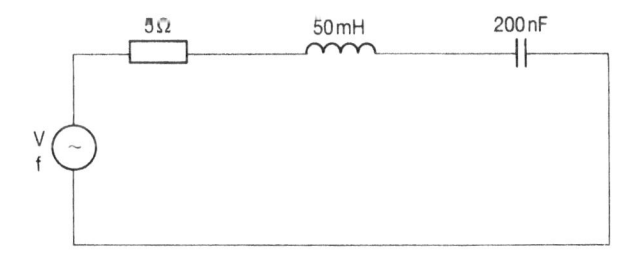

Figure 6.4

The circuit is shown in Fig. 6.4. Using Equation (6.2)

$$f_0 = 1/2\pi\sqrt{(LC)} = 1/2\pi\sqrt{(50 \times 10^{-3} \times 200 \times 10^{-9})} = 1/2\pi\sqrt{(10^{-8})}$$
$$= 1/(2\pi \times 10^{-4}) = 10^4/2\pi = 1.591 \times 10^3 \text{ Hz}$$
$$= 1.591 \text{ kHz}$$

Note that this result is independent of the resistance in the circuit, whether it be that of a separate resistor or that of the coil.

Example 6.2

Determine the value of capacitance to which the variable capacitor C must be set in order to make the circuit given in Fig. 6.5 resonate at 400 Hz.

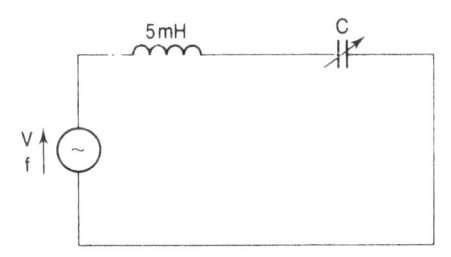

Figure 6.5

Solution

Using Equation (6.2) $f_0 = 1/2\pi\sqrt{(LC)}$ and rearranging it to make C the subject we get $C = 1/4\pi^2 f_0^2 L$ (squaring both sides and then multiplying both sides by C/f_0^2). Thus

$$C = 1/4\pi^2\ 400^2 \times 5 \times 10^{-3} = 31.6 \times 10^{-6}\ F$$

Example 6.3

A resonant series circuit consists of a capacitor having a capacitance of 0.1 μF and a coil whose inductive reactance is 60 Ω. Calculate the inductance of the coil.

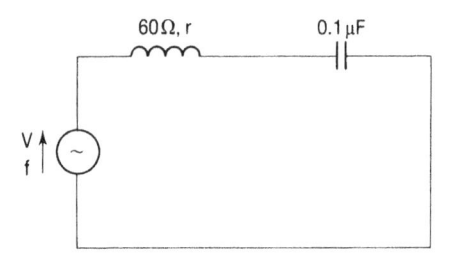

Figure 6.6

Solution

The circuit is shown in Fig. 6.6 and r is the resistance of the coil. Since the circuit is in a state of resonance we can use Equation (6.1) with $f = f_0$:

$$2\pi f_0 L = 1/2\pi f_0 C$$
$$X_L = 1/2\pi f_0 C$$
$$f_0 = 1/2\pi X_L C \quad \text{(multiplying both sides by } f_0/X_L)$$

so

$$f_0 = 1/2\pi\ 60 \times 0.1 \times 10^{-6} = 26\ 526\ Hz$$

and

$$L = X_L/2\pi f_0 = 60/2\pi \times 26\ 526 = 359.9 \times 10^{-6}\ H$$

Impedance and current at resonance

The graph of impedance $Z = \sqrt{[R^2 + (X_L - X_C)^2]}$ to a base of frequency is given in Fig. 6.7 and this shows that Z has a minimum value ($=R$) at the resonant frequency. Consequently, the circuit current at this frequency will have its maximum value ($=V/R$) as shown in Fig. 6.8.

If the resistance is small, the current could be very large and the potential difference developed across the inductance (IX_L) and the capacitance (IX_C) would then be very large (many times bigger than the supply voltage V). Great

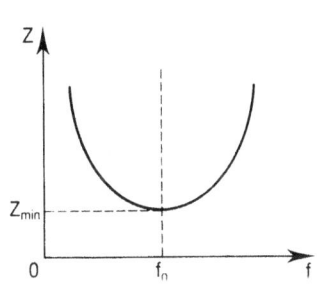

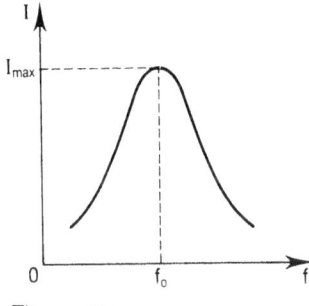

Figure 6.7

Figure 6.8

care must therefore be taken when dealing with series circuits containing inductance and capacitance.

Example 6.4

A coil having a resistance of 5 Ω and inductance of 10 mH is connected in series with a capacitor of 250 nF to a variable frequency, 100 V supply. Determine the potential difference across the capacitor at resonance.

Solution

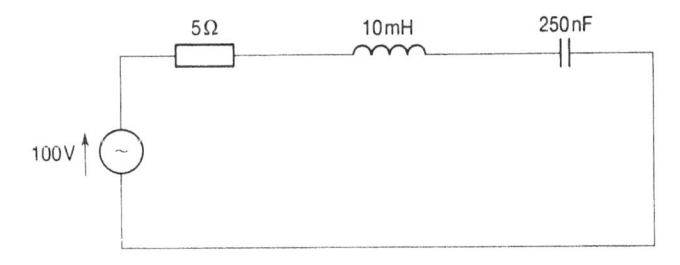

Figure 6.9

The circuit is shown in Fig. 6.9. Using Equation (6.2)

$f_0 = 1/2\pi\sqrt{[10 \times 10^{-3} \times 250 \times 10^{-9}]} = 3183$ Hz

$X_C = 1/2\pi f_0 C = 1/2\pi \times 3183 \times 250 \times 10^{-9} = 200$ Ω

The current at resonance is limited only by the resistance and $I = V/R = 100/5 = 20$ A. The potential difference across the capacitor is given by $V_C = IX_C = 20 \times 200 = 4000$ V. Note that this is 40 times greater than the supply voltage V.

Example 6.5

A 20 mH coil has a resistance of 50 Ω and is connected in series with a capacitor to a 250 mV supply. If the circuit is to resonate at 100 kHz calculate the capacitance of the capacitor and its working voltage.

Solution

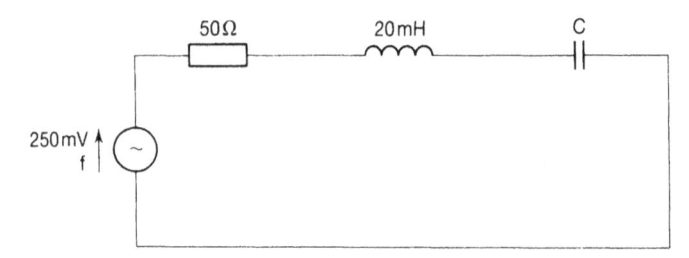

Figure 6.10

The circuit is shown in Fig. 6.10. At resonance, $f_0 = 1/2\pi\sqrt{(LC)}$ so

$C = 1/4\pi^2 f_0^2 L = 1/4\pi^2 (100 \times 10^3)^2 \, 20 \times 10^{-3} = 126.6 \times 10^{-12}$ F

Therefore

$X_C = 1/2\pi f_0 C = 1/2\pi \, 100 \times 10^3 \times 126.6 \times 10^{-12} = 12\,571 \, \Omega$

At resonance

$I = V/R = 250 \times 10^{-3}/50 = 5$ mA

The potential difference across the capacitor at resonance is

$IX_C = 5 \times 10^{-3} \times 12\,571 = 62.855$ V

The capacitor working voltage must therefore be about 65 V.

Q-factor

The fact that the circuit behaves as a pure resistor and that the power factor is unity seems to indicate that there is no reactive power in the circuit at resonance. This is not so: what is happening is that the reactive energy is continuously being transferred between the capacitor, where it is stored in the electric field, and the inductor, where it is stored in the magnetic field. Since this energy transfer is taking place within the circuit, it appears from the outside as though there is no reactive power. As the resistance of the circuit becomes smaller so the current becomes larger and the stored energy ($I^2 X_L = I^2 X_C$) oscillating between the capacitor and the inductor becomes much larger than the energy ($I^2 R$) dissipated in the resistor.

The ratio of $I^2 X_L$ to $I^2 R$ is called the Q-factor (quality factor) of the coil or circuit so that

$Q = I^2 X_L / (I^2 R) = X_L / R$

and since $X_L = 2\pi f_0 L = \omega_0 L$

$Q = \omega_0 L / R$ (6.4)

Since $\omega_0 = 1/\sqrt{(LC)}$, then $Q = L/R\sqrt{(LC)} = \sqrt{L}/(R\sqrt{C})$, so

$$Q = (1/R)[\sqrt{(L/C)}] \tag{6.5}$$

Q is the ratio of inductive reactance to resistance and since the unit of both of these is the same (the ohm), then Q itself is dimensionless.

Q-factors are of the order of 10 in the audio frequency range (up to about 20 kHz), whereas in the radio frequency range they are of the order of 10^2 and in the microwave range they can be as high as 10^3. The bigger the Q-factor the easier it is for the circuit to accept current and power at the resonant frequency so that, for example, in radio and television receivers a particular station can be selected and others (which have their own resonant frequency) can be rejected.

Remember: reducing the resistance in the circuit increases the Q-factor.

Example 6.6

The circuit shown in Fig. 6.11 operates at the resonant frequency of 11.25 kHz. Determine (1) the Q-factor of the circuit, (2) the capacitance of the capacitor, and (3) the current in the circuit.

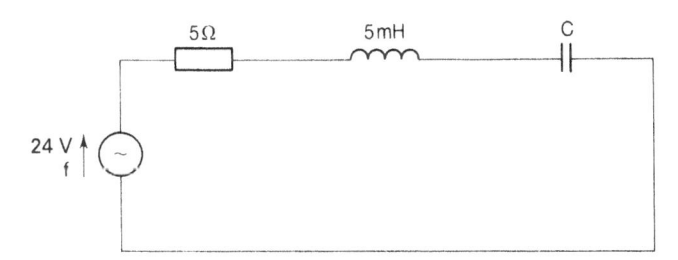

Figure 6.11

Solution

1 Using Equation (6.4) the Q-factor of the circuit is given by $Q = \omega_0 L/R$, so

$Q = 2\pi\, 11.25 \times 10^3 \times 5 \times 10^{-3}/5 = 70.68$.

2 At the resonant frequency, the capacitive reactance (X_C) is equal to the inductive reactance (X_L). The inductive reactance is
$2\pi f_0 L = 2\pi\, 11.25 \times 10^3 \times 5 \times 10^{-3} = 353.4\, \Omega$. The capacitive reactance of the capacitor is therefore also 353.4 Ω and its capacitance

$C = 1/2\pi\, 11.25 \times 10^3 \times 353.4 = 40 \times 10^{-9}$ F

3 The current is $I = V/R = 24/5 = 4.8$ A.

Filter applications

Series RLC circuits are commonly used in filter applications and a basic

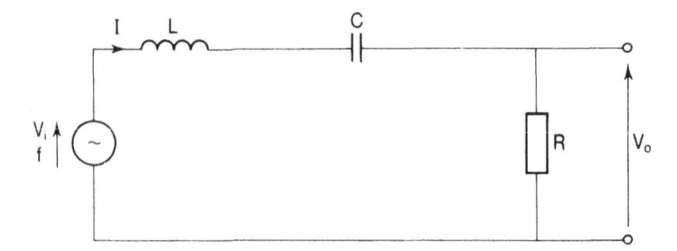

Figure 6.12

bandpass filter circuit is shown in Fig. 6.12. A bandpass filter is designed to allow signals at the resonant frequency (f_0) and those within a band of frequencies above and below f_0 to pass from the input terminals to the output terminals. Signals at frequencies outside this band are passed at a very much reduced level or not at all and are said to be rejected.

Bandwidth

For a signal to be 'passed' it has to have a voltage not less than 0.707 of the voltage at the resonant frequency f_0. The current (I), similarly, will be not less than 70.7 per cent of the current at f_0. The lowest frequency at which V_0 has this minimum value is called the lower cut-off frequency and the highest frequency at which V_0 has this minimum value is called the upper cut-off frequency. The band of frequencies which are 'passed' are those between the lower cut-off frequency and the upper cut-off frequency and this is called the pass-band. These points are illustrated in Fig. 6.13.

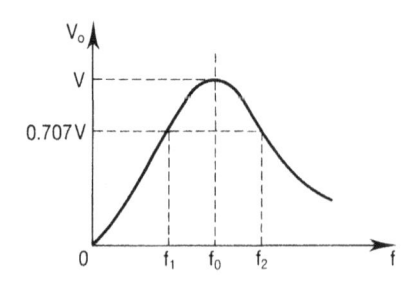

Figure 6.13

The pass-band is the band of frequencies lying between f_1 and f_2 and this is also referred to as the bandwidth (B) of the circuit so that

$$B = (f_2 - f_1) \tag{6.6}$$

The unit of bandwidth is the hertz or radians per second if angular frequency is used.

Half-power frequencies

As we have seen, at resonance the current is a maximum so that the power in the resistor is also a maximum. Now if the maximum power is $P_M = I_M^2 R$ at f_0 then at f_1 (or f_2) the power will be

$$P_1 = P_2 = (0.7071_M)^2 R = 0.5 I_M^2 R = 0.5 P_M$$

For this reason f_1 and f_2 are called the half-power frequencies.

The dB notation

The logarithm to the base ten of the ratio of two powers P_1 and P_2 is called the bel. Because this is rather large it is more usual to express power ratios in decibels (dB). This means multiplying the logarithm of the ratio by 10. By definition, a power P_2 is $10 \log_{10} (P_2/P_1)$ dB above the power P_1.

Since $P = V^2/R$, we see that a voltage V_2 is $10 \log_{10} (V_2/V_1)^2$ dB above a voltage V_1 i.e. V_2 is $20 \log_{10} (V_2/V_1)$ dB above V_1.

For the half-power frequencies f_1 and f_2, the powers (P_1 and P_2) are equal to $0.5 P_M$. Therefore P_1 is $10 \log_{10} (P_1/P_M)$ dB above P_M, i.e. P_1 is $10 \log_{10} (0.5 P_M/P_M)$ dB above P_M.

Now $10 \log_{10} (0.5) = -3$ so that P_1 is in fact -3 dB above P_M which means that it is 3 dB below P_M. It is said to be 3 dB down on P_M. For this reason the frequencies f_1 and f_2 are also referred to as the -3 dB points.

Example 6.7

The maximum current in a bandpass filter circuit is 25 mA. Calculate the current at the lower and upper cut-off frequencies.

Solution

Let the current at the lower cut-off frequency be I_1 and that at the upper cut-off frequency be I_2. Then $I_1 = I_2 = 0.707 I_M = 0.707 \times 25 = 17.68$ mA.

Example 6.8

A bandpass filter circuit has a lower cut-off frequency (f_1) = 12 kHz and an upper cut-off frequency (f_2) = 18 kHz. Calculate the bandwidth of this circuit.

Solution

Using Equation (6.6) the bandwidth is $B = f_2 - f_1 = (18 - 12)$ kHz = 6 kHz.

Gain and phase diagrams

In the circuit of Fig. 6.12 the ratio V_o/V_i is called the gain ratio and is denoted by H. Since $V_o = IR$ and $V_i = I(R + j\omega L + 1/j\omega C)$ then

$$H = IR/I(R + j\omega L + 1/j\omega C) \tag{6.7}$$

132 Resonance

Dividing both numerator and denominator by IR we have

$$H = 1/[1 + j\omega L/R + 1/j\omega CR]$$

$$H = 1/\{1 + j[\omega L/R - 1/\omega CR]\} \tag{6.8}$$

Now from Equation (6.4), $Q = \omega_0 L/R$ and multiplying both sides by ω/ω_0 we get

$$(\omega/\omega_0)Q = \omega L/R \tag{6.9}$$

Also, since $\omega_0 L/R = 1/\omega_0 CR$ (because $\omega_0 L = 1/\omega_0 C$ at resonance then)

$$Q = 1/\omega_0 RC \tag{6.10}$$

Multiplying both sides by ω_0/ω we have

$$(\omega_0/\omega)Q = 1/\omega RC \tag{6.11}$$

Substituting from Equations (6.9) and (6.11) into Equation (6.8) we see that

$$H = 1/\{1 + jQ[(\omega/\omega_0) - (\omega_0/\omega)]\} \tag{6.12}$$

The phase angle is given by

$$\tan^{-1}[(\omega L - 1/\omega C)/R] \tag{6.13}$$

If H is plotted to a base of frequency we obtain the graph in Fig. 6.14 and the graph of the phase angle to the same base is given in Fig. 6.15. Note that when $X_L = X_C$, $\phi = 0$; when $X_L = 0$, ϕ lies between 0 and $-90°$; when $X_C = 0$, ϕ lies between 0 and $+90°$.

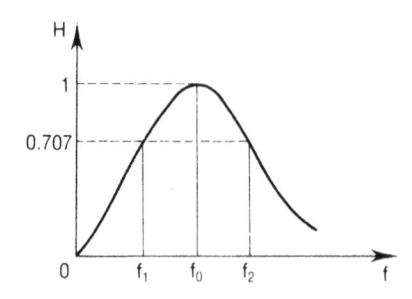

Figure 6.14

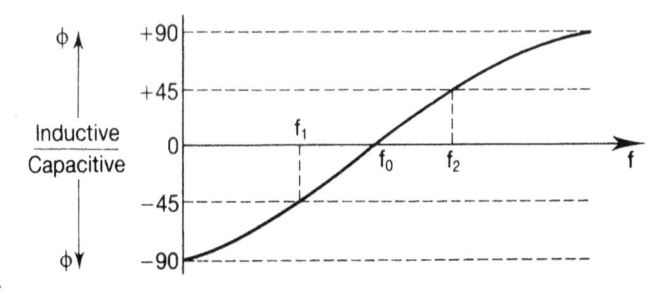

Figure 6.15

6.2 PARALLEL RESONANCE

Also in Chapter 4 we saw that under certain conditions the parallel RLC circuit behaves as a pure resistor because the phasors representing the circuit supply voltage and the total current drawn from the supply are in phase with each other. This condition is known as parallel resonance and in order to analyse it more fully the circuit diagram is given again in Fig. 6.16, the relevant phasor diagram being shown in Fig. 6.17.

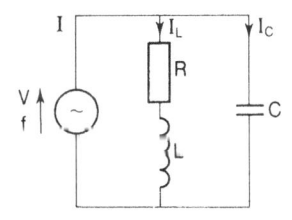

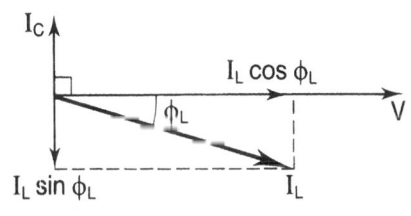

Figure 6.16

Figure 6.17

The important relationships in the circuit of Fig. 6.16 are:

$$I = I_C + I_L \quad \text{phasorially} \tag{6.14}$$

$$I_C = V/X_C \tag{6.15}$$

$$I_L = V/Z_L \tag{6.16}$$

where $Z_L = R + jX_L = \sqrt{(R^2 + X_L^2)}$.

The phasor diagram of Fig. 6.17 is drawn with the supply voltage V as the reference phasor. The current phasor I_C is drawn leading V by 90°; the current phasor I_L is drawn lagging V by an angle ϕ_L where $\phi_L = \tan^{-1}(X_L/R)$ and is the phase angle of the branch containing the inductance.

From this phasor diagram it is seen that for the circuit supply voltage (V) and the total current drawn from the supply (I) to be in phase,

$$I_C = I_L \sin \phi_L \tag{6.17}$$

where I_C is the current through the capacitor C, I_L is the current through the inductance L, ϕ_L is the phase angle of the branch containing R and L, and

$$\phi_L = \sin^{-1}(X_L/Z_L) \tag{6.18}$$

From Equations (6.15), (6.16) and (6.18) we have, by substitution in Equation (6.17),

$$V/X_C = VX_L/[\sqrt{(R^2 + X_L^2)} \times \sqrt{(R^2 + X_L^2)}]$$

Dividing throughout by V and putting $X_C = 1/2\pi f_0 C$ and $X_L = 2\pi f_0 L$ (f_0 being the resonant frequency) we have

$$2\pi f_0 C = 2\pi f_0 L / [R^2 + (2\pi f_0 L)^2]$$

Multiplying both sides by $[R^2 + (2\pi f_0 L)^2]/2\pi f_0 C$ we get

$$R^2 + (2\pi f_0 L)^2 = L/C \qquad (6.19)$$

$$(2\pi f_0 L)^2 = (L/C) - R^2$$
$$(2\pi f_0)^2 = (1/LC) - (R/L)^2 \quad \text{(dividing both sides by } L^2\text{)}$$
$$2\pi f_0 = \sqrt{[(1/LC) - (R/L)^2]}$$
$$f_0 = (1/2\pi)\sqrt{[(1/LC) - (R/L)^2]} \qquad (6.20)$$

If the resistance (R) is very much smaller than the inductive reactance ($2\pi f_0 L$), which is normally the case, then Equation (6.19) becomes $(2\pi f_0 L)^2 = L/C$ so that

$$(2\pi f_0)^2 = 1/LC \quad \text{(dividing both sides by } 1/L^2\text{)}$$
$$2\pi f_0 = \sqrt{(1/LC)}$$
$$f_0 = 1/2\pi\sqrt{(LC)} \qquad (6.21)$$

Compare this with the Equation (6.2) for the resonant frequency of a series RLC circuit.

Dynamic impedance

Remembering that the equivalent impedance (Z_{eq}) of a parallel combination of two impedances Z_1 and Z_2 is given by $Z_{eq} = Z_1 Z_2/(Z_1 + Z_2)$ we see that the general expression for the impedance of the parallel RLC circuit of Fig. 6.16 is

$$Z = (R + j\omega L)(-j/\omega C)/[(R + j\omega L) + (-j/\omega C)]$$

Dividing the numerator and denominator by $(-j/\omega C)$ we have

$$Z = (R + j\omega L)/[(R/-j/\omega C) + (j\omega L/-j/\omega C) + (-j/\omega C/-j/\omega C)]$$
$$= (R + j\omega L)/[j\omega CR - \omega^2 LC + 1]$$

Again, assuming that $\omega L \gg R$, this reduces to

$$Z = j\omega L/[j\omega CR + (1 - \omega^2 LC)] \qquad (6.22)$$

At resonance, $\omega^2 = \omega_0^2 = 1/LC$ and then

$$Z = j\omega_0 L/[j\omega_0 CR + (1 - 1)] = L/CR$$

This has the characteristics of a pure resistance and is called the dynamic impedance (Z_d) of the tuned circuit. Thus

$$Z_d = L/CR \qquad (6.23)$$

Just as in the case of the series tuned circuit, in this parallel circuit the ratio $\omega_0 L/R$ is called the Q-factor of the circuit. In this case, though, as Z_d becomes larger, so the current becomes smaller and the Q-factor becomes larger.

Example 6.9

A coil having an inductance of 200 μH and a resistance of 50 Ω is connected in parallel with a capacitor having a capacitance of 120 pF to a 100 V supply. Determine (1) the resonant frequency, (2) the dynamic impedance of the circuit, (3) the Q-factor of the circuit, and (4) the current in the circuit at resonance.

Solution

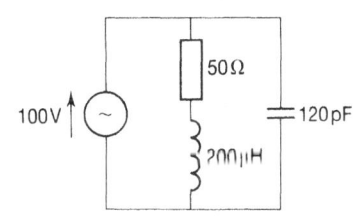

Figure 6.18

The circuit is shown in Fig. 6.18.

1 From Equation (6.20),

$$f_0 = (1/2\pi)\sqrt{[(1/LC) - (R/L)^2]}$$
$$= (1/2\pi)\sqrt{\{[(1/(200 \times 10^{-6} \times 120 \times 10^{-12})] - (50/200 \times 10^{-6})^2\}}$$
$$= (1/2\pi)\sqrt{(4166 \times 10^{10} - 6.25 \times 10^{10})}$$
$$= 1.02 \text{ MHz}$$

Note that $(R/L)^2 \ll 1/LC$.

2 From Equation (6.23),
$$Z_d = L/CR = 200 \times 10^{-6}/(120 \times 10^{-12} \times 50) = 33.3 \text{ k}\Omega$$

3 The Q-factor is given by
$$\omega_0 L/R = 2\pi \times 1.02 \times 10^6 \times 200 \times 10^{-6}/50 = 25.64$$

4 At resonance, the current is limited by the dynamic impedance so
$$I = V/Z_d = 100/(33.3 \times 10^3) = 3 \text{ mA}$$

Bandwidth

We have seen that, assuming $R \ll X_L$, the impedance of the circuit of Fig. 6.16 is given by Equation (6.22) to be

$$Z = j\omega L/[j\omega CR + (1 - \omega^2 LC)]$$

Factorizing the denominator we get

$$Z = j\omega L/\{j\omega CR[1 + (1/j\omega CR) - (\omega^2 LC/j\omega CR)]\}$$
$$= j\omega L/\{j\omega CR[1 + (LC/j\omega CR)((1/LC) - \omega^2)]\}$$

Dividing the numerator and the denominator by $j\omega CR$ this becomes

$$Z = (L/CR)/[1 + (LC/j\omega CR)\{(1/LC) - \omega^2\}]] \qquad (6.24)$$

Now at resonance, $L/CR = Z_d$, the dynamic impedance; $1/\omega CR\ (=\omega L/R) = Q$; and $LC = 1/\omega_0^2$. Substituting these into Equation (6.24) we obtain

$$Z = Z_d/[1 + (Q/j\omega_0^2)(\omega_0^2 - \omega^2)]$$

Rearranging to put the 'j' in the numerator we have

$$Z = Z_d/[1 + (jQ/j\omega_0^2)(\omega^2 - \omega_0^2)] \qquad (6.25)$$

For frequencies close to resonance, we could say that $\omega = \omega_0 + \delta\omega$ where $\delta\omega$ is a small frequency deviation. In that case

$$\begin{aligned}(\omega^2 - \omega_0^2) &= (\omega_0 + \delta\omega)^2 - \omega_0^2 \\ &= \omega_0^2 + 2\omega_0\delta\omega + \delta\omega^2 - \omega_0^2 \\ &= 2\omega_0\delta\omega + \delta\omega^2 \\ &= 2\omega_0\delta\omega \quad \text{(since } \delta\omega^2 \to 0\text{)}\end{aligned}$$

Substituting this in Equation (6.25) we have

$$Z = Z_d/[1 + (jQ/\omega_0^2)(2\omega_0\delta\omega)] = Z_d/[1 + j2Q(\delta\omega/\omega_0)]$$

The magnitude of this is

$$Z = Z_d/\sqrt{\{1 + [2Q(\delta\omega/\omega_0)]^2\}} \qquad (6.26)$$

The bandwidth of a parallel resonant circuit is defined to be the frequency range between the two frequencies for which $Z/Z_d = 1/\sqrt{2}$, and from Equation (6.26) $Z/Z_d = 1/\sqrt{\{1 + [2Q(\delta\omega/\omega_0)]^2\}}$, so to obtain the bandwidth we put $1/\sqrt{\{1 + [2Q(\delta\omega/\omega_0)]^2\}} = 1/\sqrt{2}$. Thus, inverting both sides and taking the square root, we get

$$\begin{aligned}1 + [2Q(\delta\omega/\omega_0)]^2 &= 2 \\ [2Q(\delta\omega/\omega_0)]^2 &= 1 \\ 2Q(\delta\omega/\omega_0) &= 1 \\ \delta\omega/\omega_0 &= 1/2Q \\ \delta\omega &= \omega_0/2Q\end{aligned}$$

The bandwidth is the band of frequencies from $-\delta\omega$ to $+\delta\omega$ (i.e. $2\delta\omega$). Thus

$$B = 2\delta\omega = 2\omega_0/2Q = \omega_0/Q$$

Finally

$$B = \omega_0/Q \qquad (6.27)$$

Example 6.10

Determine (1) the bandwidth of the circuit in Example 6.9, (2) the effect on the bandwidth of this circuit of adding a resistance of 50 Ω in series with the coil.

Solution

1 From Equation (6.27) the bandwidth is given by
  $B = \omega_0/Q$ (rad s^{-1}) $= f_0/Q$(Hertz). Therefore
  $B = f_0/Q = 1.02/25.64 = 0.04$ MHz

2 Note that the addition of 50 Ω in the inductive branch has negligible effect on the resonant frequency so ω_0 is unchanged. This is because, as we saw in Example 6.9, $1/LC \gg (R/L)^2$. Since $Q = \omega_0 L/R$, then with ω_0 and L unchanged, the effect of doubling R is to halve Q and to double B. The bandwidth is therefore doubled.

6.3 SELF-ASSESSMENT TEST

1 State the condition for a series circuit containing inductance and capacitance to be in a state of resonance.

2 Give an expression for the resonant frequency of a series RLC circuit.

3 What is the power factor of a series RLC circuit when it is in a state of resonance?

4 A series circuit has an inductance of 2 H and a capacitance of 8 μF. What is its resonant angular frequency?

5 A series circuit has a resistance of 2 Ω, an inductance of 10 mH and a capacitance of 0.1 μF. What is the value of the impedance of this circuit at resonance?

6 Explain why the voltage developed across the inductor and capacitor of a series resonant circuit could be many times greater than the supply voltage.

7 Define the Q-factor of a coil or circuit.

8 Give the unit of Q.

9 How may the Q-factor of a circuit be increased?

10 Explain what is meant by a 'bandpass filter'.

11 Define the bandwidth of a circuit.

12 Give the unit of bandwidth.

13 Explain the meaning of 'half-power frequency'.

14 Give the relationship between two powers P_1 and P_2 in decibel form.

15 State the condition for a parallel circuit containing inductance and capacitance to be in a state of resonance.

138 Resonance

16 Give an expression for the resonant frequency of a parallel circuit assuming that the resistance in the circuit is very much smaller than the inductive reactance.

17 Give an expression for the dynamic impedance of a parallel circuit.

18 State what happens to the Q-factor of a parallel circuit as its dynamic impedance becomes larger.

19 State whether the current in a parallel circuit at resonance is a maximum or a minimum.

20 Give an expression for the bandwidth (B) in terms of the angular resonant frequency (ω_0) and the Q-factor.

6.4 PROBLEMS

1 A coil is connected in series with a capacitor of 20 μF to a 200 V variable frequency supply. The current is a maximum at 50 A when the frequency is set to 50 Hz. Determine the resistance and inductance of the coil.

2 A coil having an inductance of 1 H and a resistance of 5 Ω is connected in series with a resistance of 5 Ω and a capacitor of 15.8 μF. The whole combination is connected to a 200 V variable frequency supply. Determine (a) the resonant frequency, (b) the current in the circuit at resonance, (c) the corresponding voltage developed across the capacitor.

3 A series circuit consists of a 40 Ω resistor, a 0.5 H inductor and a variable capacitor connected across a 100 V, 50 Hz supply. Calculate (a) the value of the capacitance required to give resonance, (b) the voltages across the resistor, the inductor and the capacitor at resonance, and (c) the Q-factor of the circuit.

4 A bandpass filter consists of a capacitor of 5 nF in series with a coil of inductance 10 mH having a resistance 5 Ω and a resistor of 75 Ω resistance. The output voltage is taken across the resistor. Determine (a) the resonant frequency, (b) the Q-factor of the coil and (c) the bandwidth of the filter.

5 Determine the voltage gain as a ratio (V_o/V_{in}) and in decibels of the circuit of Problem 4 at 20 kHz.

6 A voltage V_i is applied to a circuit consisting of a capacitor in series with a resistor. An output voltage V_o is taken across the resistor (this constitutes a simple high-pass filter circuit).
   (a) Obtain an expression for the voltage gain $H(j\omega)$.
   (b) If $1/CR = 1000$ obtain the voltage ratio in dB for $\omega = 10$; $\omega = 100$ and $\omega = 1000$ rad^{-1}. Comment on the result.
   (c) State how the circuit could be converted into a bandpass filter circuit.

7 A circuit consisting of a coil of inductance 250 mH, having a resistance of 20 Ω, in parallel with a variable capacitor C is connected to a 200 V 50 Hz supply. Determine (a) the value of C required for the circuit to resonate, (b) the power absorbed at resonance, and (c) the ratio of the current through the capacitor to the supply current at resonance.

8 A resistor of 90 Ω resistance is connected in series with a coil of inductance 500 mH, having a resistance of 10 Ω. This series circuit is connected in parallel with a 20 μF capacitance across a 250 V variable frequency supply. Determine (a) the resonant frequency of the circuit, (b) the resonant frequency if the 90 Ω resistor is short circuited and (c) the current drawn from the supply in each case.

9 A coil of inductance 10 mH and resistance 50 Ω is connected in parallel with a capacitor of 0.01 μF. Determine (a) the resonant frequency, (b) the Q-factor, (c) the bandwidth of the circuit, and (d) the half-power frequencies.

10 A series circuit consisting of an inductance of 0.3 H, having a resistance of 10 Ω, and a variable capacitor C_1 is supplied from a 100 V, variable frequency, source.
  (a) Determine the value of C_1 necessary for the circuit to operate resonantly at 50 Hz.
  (b) A second variable capacitor, C_2, is now connected in parallel with the original circuit and the supply frequency is adjusted to 60 Hz. Determine the value of C_2 in order that the circuit still operates with minimum current.

11 A coil of resistance 2 Ω has a Q-factor of 80 and is to work at a frequency of 1 kHz. Determine
  (a) the value of the equivalent parallel resistance of the coil for the same Q-factor,
  (b) the additional parallel resistance required for the Q-factor to be halved at the same frequency,
  (c) the capacitance required to give the circuit a dynamic impedance of 100 kΩ, and
  (d) the values of the Q-factor and frequency under condition (c).

12 A coil having a Q-factor of 100 is connected in parallel with a capacitor of 100 pF. The circuit resonates at a frequency of 5 MHz. Determine (a) the bandwidth of the circuit, (b) the amount of resistance required to be placed in parallel with the capacitor in order to increase the bandwidth to 250 kHz, and (c) the amount of resistance required to be placed in series with the inductor in order to produce the same bandwidth.

13 The aerial circuit of a radio receiver consists of a tuned circuit comprising a coil of inductance 100 μH and resistance 5 Ω in parallel with a variable

capacitor. A resistor of 2 kΩ is connected in series with this parallel circuit and the whole combination is connected to a 10 V, 300 kHz supply. Calculate (a) the value of the capacitance required to give resonance at 300 kHz, (b) the dynamic impedance of the circuit, (c) the Q-factor, and (d) the current through the capacitor at resonance. Show that the ratio of the capacitor current to the total current is equal to the Q-factor.

7 Nodal and mesh analysis

7.1 INTRODUCTION

For circuits which are more complicated than those considered in the previous chapters, it is still possible to analyse them using the methods described there but the working can become extremely tedious. Two of the methods devised to make things more manageable are the nodal voltage and the mesh current approaches. These lend themselves to matrix methods of solution, both manually and by computer. We shall begin therefore by setting out the relevant parts of the matrix algebra techniques.

7.2 MATRICES

A matrix is a rectangular array of numbers (or letters or functions) arranged in rows and columns. The individual numbers are called elements of the matrix and these are often identified by the use of double subscripts, the first part indicating the row and the second indicating the column in which the element is situated. Thus, for example, element a_{34} is an element in the 4th column of the 3rd row. Matrices are usually enclosed by square brackets so that

$$\begin{bmatrix} 2 & 5 & 9 & 8 \\ -4 & 6 & -1 & 10 \end{bmatrix}$$

is a matrix having two rows and four columns. It is said to be a matrix of order 2×4. In this matrix, element $a_{13} = 9$ and element $a_{21} = -4$.

Example 7.1

Write down the matrix of order 4×2 for which $a_{11} = 1$; $a_{12} = 0$; $a_{21} = -3$; $a_{22} = 4$; $a_{31} = -6$; $a_{32} = 9$; $a_{41} = 0$ and $a_{42} = 5$

Solution

A matrix of order 4×2 has four rows and two columns and in this case the elements are defined numerically so we may write

$$\begin{bmatrix} 1 & 0 \\ -3 & 4 \\ -6 & 9 \\ 0 & 5 \end{bmatrix}$$

A row matrix

A row matrix has only one row. Thus $[X \ Y \ Z]$ is a three-column row matrix. A row matrix is also called a row vector.

A column matrix

A column matrix has only one column. Thus

$$\begin{bmatrix} 1 \\ 3 \\ 5 \\ 9 \end{bmatrix}$$

is a four-row column matrix. A column matrix is also called a column vector.

Matrix addition and subtraction

This is only possible for matrices of the same order and is carried out by adding or subtracting corresponding elements of the matrices being added or subtracted.

Example 7.2

If

$$A = \begin{bmatrix} 4 & 6 & Z \\ Y & 2 & -1 \end{bmatrix} \quad \text{and} \quad B = \begin{bmatrix} 3 & X & 2 \\ -Y & 4 & 6 \end{bmatrix}$$

obtain (1) $A + B$, (2) $A - B$.

Solution

$$1 \ A + B = \begin{bmatrix} (4+3) & (6+X) & (Z+2) \\ Y+(-Y) & (2+4) & (-1+6) \end{bmatrix} = \begin{bmatrix} 7 & (6+X) & (Z+2) \\ 0 & 6 & 5 \end{bmatrix}$$

$$2 \ A - B = \begin{bmatrix} (4-3) & (6-X) & (Z-2) \\ Y-(-Y) & (2-4) & (-1-6) \end{bmatrix} = \begin{bmatrix} 1 & (6-X) & (Z-2) \\ 2Y & -2 & -7 \end{bmatrix}$$

Note that $A + B = B + A$ so that the process is commutative. It is also associative, so that for the addition of three matrices A, B and C, $A + (B + C) = (A + B) + C$.

Multiplication of matrices

Multiplication of two matrices is only possible if the number of rows in one of the matrices is equal to the number of columns in the other. If two matrices (A and B) are multiplied to give a third matrix (C) then any element (c_{mn}) of C is found by adding the products of all the elements in row m of A with the corresponding elements in column n of B.

Example 7.3

If

$$A = \begin{bmatrix} 3 & 2 \\ 5 & 1 \end{bmatrix} \quad \text{and} \quad B = \begin{bmatrix} 9 & 4 \\ 6 & 7 \end{bmatrix}$$

obtain (1) AB, (2) BA.

Solution

1 Let $C = AB$. Then

$$C = \begin{bmatrix} 3 & 2 \\ 5 & 1 \end{bmatrix}\begin{bmatrix} 9 & 4 \\ 6 & 7 \end{bmatrix} = \begin{bmatrix} (3 \times 9) + (2 \times 6) & (3 \times 4) + (2 \times 7) \\ (5 \times 9) + (1 \times 6) & (5 \times 4) + (1 \times 7) \end{bmatrix} = \begin{bmatrix} 39 & 26 \\ 51 & 27 \end{bmatrix}$$

2 Let $D = BA$. Then

$$D = \begin{bmatrix} 9 & 4 \\ 6 & 7 \end{bmatrix}\begin{bmatrix} 3 & 2 \\ 5 & 1 \end{bmatrix} = \begin{bmatrix} (9 \times 3) + (4 \times 5) & (9 \times 2) + (4 \times 1) \\ (6 \times 3) + (7 \times 5) & (6 \times 2) + (7 \times 1) \end{bmatrix} = \begin{bmatrix} 47 & 22 \\ 53 & 19 \end{bmatrix}$$

Note that $C \neq D$ so that $AB \neq BA$. The multiplication process is therefore not commutative and care must be taken to multiply in the correct order.

The determinant of a matrix

The determinant of a matrix is a number which is obtained by subtracting the sum of the products of the elements along the diagonals to the left from the sum of the products of the elements along the diagonals to the right. The symbol for the determinant is Δ and the elements are enclosed by vertical lines.

Example 7.4

Find the determinant of the matrix

$$\begin{bmatrix} 3 & 2 \\ 5 & 1 \end{bmatrix}$$

Solution

$$\Delta = \begin{vmatrix} 3 & 2 \\ 5 & 1 \end{vmatrix}$$

The product of the elements along the diagonal to the right is $3 \times 1 = 3$.
The product of the elements along the diagonal to the left is $2 \times 5 = 10$.
Therefore $\Delta = 3 - 10 = -7$.

Example 7.5

Find the determinant of the 3×3 matrix

$$\begin{bmatrix} 4 & 2 & 1 \\ 2 & 4 & 3 \\ 5 & 0 & 6 \end{bmatrix}$$

Solution

$$\Delta = \begin{vmatrix} 4 & 2 & 1 \\ 2 & 4 & 3 \\ 5 & 0 & 6 \end{vmatrix}$$

The sum of the products of the elements along the diagonals to the right is

$$(4 \times 4 \times 6) + (2 \times 3 \times 5) + (1 \times 2 \times 0) = 126$$

The sum of the products of the elements along the diagonals to the left is

$$(1 \times 4 \times 5) + (2 \times 2 \times 6) + (4 \times 3 \times 0) = 44$$

Therefore $\Delta = 126 - 44 = 82$.

It might be found helpful to set out the elements of the determinant again alongside the original one to see the three diagonals in each direction.

The minor of an element

This is defined as the determinant of the submatrix obtained by deleting the row and the column containing the element. Thus for the matrix

$$\begin{bmatrix} a_{11} & a_{12} & a_{13} \\ a_{21} & a_{22} & a_{23} \\ a_{31} & a_{32} & a_{33} \end{bmatrix}$$

the minor of element a_{21} is

$$\begin{vmatrix} a_{12} & a_{13} \\ a_{32} & a_{33} \end{vmatrix}$$

Cofactors

The cofactor of an element a_{mn} in a square matrix is defined to be $(-1)^{m+n}$ times the minor of the element.

Example 7.6

Find the cofactors of the elements in row 1 of the matrix

$$A = \begin{bmatrix} 1 & 3 & 4 \\ 2 & 2 & 3 \\ 1 & 1 & -2 \end{bmatrix}$$

Solution

The cofactor of element a_{11} is found by removing the first row and the first column and multiplying the determinant of the resulting 2×2 submatrix (i.e. the minor of the element) by $(-1)^{(1+1)}$. Thus the cofactor of the element a_{11} is

$$(-1)^{(1+1)} \begin{vmatrix} 2 & 3 \\ 1 & -2 \end{vmatrix} = (-1)^2[(2 \times -2) - (3 \times 1)] = -7$$

Similarly, by removing the first row and the second column we find the cofactor of a_{12}. Therefore the cofactor of the element a_{12} is

$$(-1)^{(1+2)} \begin{vmatrix} 2 & 3 \\ 1 & -2 \end{vmatrix} = (-1)^3[(2 \times -2) - (3 \times 1)] = 7$$

We remove the first row and the third column to find the cofactor of the element a_{13}. Therefore the cofactor of the element a_{13} is

$$(-1)^{(1+3)} \begin{vmatrix} 2 & 2 \\ 1 & 1 \end{vmatrix} = (-1)^4[(2 \times 1) - (2 \times 1)] = 0$$

Evaluation of Δ using cofactors

Determinants can be calculated in terms of cofactors using the rule

$$\Delta = \sum_{m=1}^{x} a_{mn} c_{mn} \qquad (7.1)$$

where x is the order of the square matrix, a_{mn} is the element, and c_{mn} is the cofactor of the element.

Example 7.7

Evaluate the determinant of the matrix

146 Nodal and mesh analysis

$$\begin{bmatrix} 2 & -2 & 4 \\ 0 & 4 & -2 \\ 4 & -1 & -1 \end{bmatrix}$$

Solution

This is a square matrix of order 3. The elements in column 1 ($n = 1$) are $a_{11} = 2$; $a_{21} = 0$; $a_{31} = 4$. From Equation (7.1), taking $n = 1$,

$$\Delta = \sum_{m=1}^{3} a_{m1}c_{m1} = a_{11}c_{11} + a_{21}c_{21} + a_{31}c_{31}$$

Now c_{11} is the cofactor of element a_{11} and by definition

$$c_{11} = (-1)^{(1+1)} \begin{vmatrix} a_{22} & a_{23} \\ a_{32} & a_{33} \end{vmatrix}$$

Similarly

$$c_{21} = (-1)^{(2+1)} \begin{vmatrix} a_{12} & a_{13} \\ a_{32} & a_{33} \end{vmatrix} \quad \text{and} \quad c_{31} = (-1)^{(3+1)} \begin{vmatrix} a_{12} & a_{13} \\ a_{22} & a_{23} \end{vmatrix}$$

Therefore

$$\Delta = 2(-1)^2 \begin{vmatrix} 4 & -2 \\ -1 & -1 \end{vmatrix} + (0 \times c_{21}) + 4(-1)^4 \begin{vmatrix} -2 & 4 \\ 4 & -2 \end{vmatrix} = (2)[-4 - 2] + (4)[4 - 16]$$

$$= -12 - 48 = -60$$

Cramer's rule

This is most useful for solving simultaneous equations using matrix methods. A set of simultaneous equations with unknowns in x may be written

$$Ax = B$$

where A is a rectangular matrix of the coefficients of x and B is a column matrix. Cramer's rule states that to find x_m we obtain the determinant Δ of A and divide it into Δ_m, where Δ_m is the determinant of the matrix obtained by replacing the mth column of A by the column matrix B.

Example 7.8

The three mesh equations of a certain circuit are

$$5I_1 + 2I_2 + I_3 = 5$$
$$I_1 + 10I_2 - 2I_3 = 10$$
$$2I_1 - 3I_2 + 4I_3 = 0$$

Determine the current I_1.

Solution

In matrix form the equations may be written $[A][I] = [B]$ where

$$A = \begin{bmatrix} 5 & 2 & 1 \\ 1 & 10 & -2 \\ 2 & -3 & 4 \end{bmatrix} \text{ and } B = \begin{bmatrix} 5 \\ 10 \\ 0 \end{bmatrix}$$

Using Cramer's rule we have $I_1 = \Delta_1/\Delta$.

Δ is the determinant of A and

$$\Delta = \begin{vmatrix} 5 & 2 & 1 \\ 1 & 10 & -2 \\ 2 & -3 & 4 \end{vmatrix}$$

The sum of the products of the elements along the diagonals to the right is

$(5 \times 10 \times 4) + (2 \times -2 \times 2) + (1 \times 1 \times -3) = 200 - 8 - 3 = 189$

The sum of the products of the elements along the diagonals to the left is

$(1 \times 10 \times 2) + (2 \times 1 \times 4) + (5 \times -2 \times -3) = 20 + 8 + 30 = 58$

Therefore

$\Delta = 189 - 58 = 131$

To find Δ_1 we replace the first column of matrix A with the column matrix B and calculate the determinant of the resulting matrix. Thus

$$\Delta_1 = \begin{vmatrix} 5 & 2 & 1 \\ 10 & 10 & -2 \\ 0 & -3 & 4 \end{vmatrix}$$

The sum of the products of the elements along the diagonals to the right is

$(5 \times 10 \times 4) + (2 \times -2 \times 0) + (1 \times 10 \times -3) = 200 - 30 = 170$

The sum of the products of the elements along the diagonals to the left is

$(1 \times 10 \times 0) + (2 \times 10 \times 4) + (5 \times -2 \times -3) = 80 + 30 = 110$

Therefore

$\Delta_1 = 170 - 110 = 60$

Finally, $I_1 = \Delta_1/\Delta = 60/131 = 0.458$ A

7.3 NODAL VOLTAGE ANALYSIS

This method of circuit analysis involves identifying the nodes in a circuit, selecting one of them as the reference node and then referring the voltages at

148 Nodal and mesh analysis

all the other nodes to it. By applying Kirchhoff's current law to each of these other nodes in turn, a set of equations can be obtained from which the various nodal voltages may be calculated. The node chosen as the reference can be purely arbitrary but will normally be the one connected to an earthed or grounded part of the circuit. Otherwise the node at the bottom of the circuit is usually chosen.

To illustrate the method, first consider the circuit shown in Fig. 7.1.

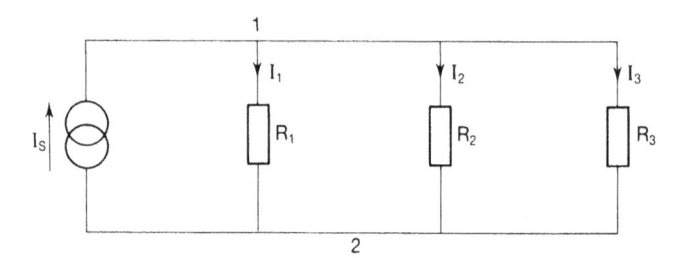

Figure 7.1

This circuit has just two nodes which are identified as 1 and 2. Let the voltage of nodes 1 and 2 be V_1 and V_2, respectively. Applying KCL to node 1 we have

$$I_S = I_1 + I_2 + I_3 = (V_1 - V_2)/R_1 + (V_1 - V_2)/R_2 + (V_1 - V_2)/R_3$$

We choose the reference node to be node 2 and, since only potential differences are important, we can make $V_2 = 0$:

$$I_S = V_1[1/R_1 + 1/R_2 + 1/R_3]$$

Using conductances $(G = 1/R)$ we have

$$I_S = V_1[G_1 + G_2 + G_3]$$

Now let us consider the slightly more complicated circuit of Fig. 7.2 which has three nodes. The three nodes are identified as 1, 2 and 3 and we choose node 3 as the reference. Let the node voltages be V_1, V_2 and V_3. Applying KCL to node 1:

$$I = I_1 + I_2 + I_3 = (V_1 - V_3)/R_1 + (V_1 - V_3)/R_2 + (V_1 - V_2)/R_3$$

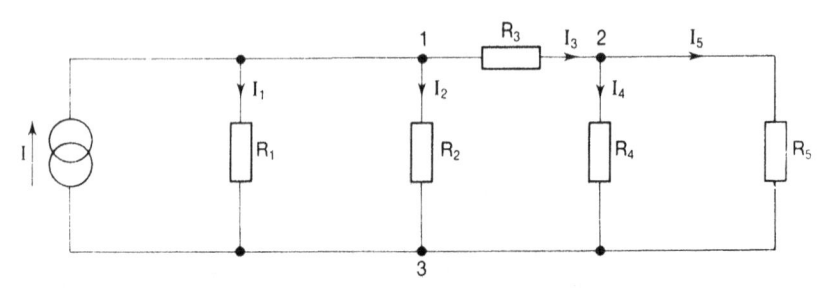

Figure 7.2

7.3 Nodal voltage analysis

Setting the reference voltage to zero, $V_3 = 0$, and then we have

$$I = V_1/R_1 + V_1/R_2 + (V_1 - V_2)/R_3$$

Using conductances

$$I = G_1V_1 + G_2V_2 + G_3(V_1 - V_2)$$

Finally

$$(G_1 + G_2 + G_3)V_1 - G_3V_2 = I \qquad (7.2)$$

Applying KCL to node 2:

$$I_3 = I_4 + I_5$$
$$(V_1 - V_2)/R_3 = (V_2 - V_3)/R_4 + (V_2 - V_3)/R_5$$

It is vitally important to get the signs correct here. For a current leaving a node, the node voltage is positive with respect to the other node. Thus we write $(V_2 - V_3)/R_4$ for I_4 which *leaves* node 2 and $(V_2 - V_3)/R_5$ for I_5 which *leaves* node 2. However, for I_3 which *enters* node 2 we must write $(V_1 - V_2)/R_3$.

Using conductances and putting $V_3 = 0$ we get

$$G_3(V_1 - V_2) = G_4V_2 + G_5V_2$$

Rearranging, we have

$$(G_3 + G_4 + G_5)V_2 - G_3V_1 = 0 \qquad (7.3)$$

We see from Equations (7.2) and (7.3) that for node 1:

- the coefficient of V_1 is the sum of all the conductances connected to node 1;
- the coefficient of V_2 is (-1) times the conductance connected between node 2 and node 1;
- the right-hand side of the equation is the current flowing into the node from the source I;

and for node 2:

- the coefficient of V_2 is the sum of all the conductances connected to node 2;
- the coefficient of V_1 is (-1) times the conductance connected between node 1 and node 2;
- the right-hand side of the equation is zero because there is no source feeding current directly into node 2.

In matrix form Equations (7.2) and (7.3) may be written

$$\begin{bmatrix} (G_1 + G_2 + G_3) & -G_3 \\ -G_3 & (G_3 + G_4 + G_5) \end{bmatrix} \begin{bmatrix} V_1 \\ V_2 \end{bmatrix} = \begin{bmatrix} I \\ 0 \end{bmatrix} \qquad (7.4)$$

150 Nodal and mesh analysis

This is of the form $Ax = B$ so that Cramer's rule may be used to solve for V_1 and V_2.

Circuits with voltage sources

The circuits considered in the previous examples have had current sources only. If a voltage source exists it could be replaced by its equivalent current source and then the analysis proceeds as described above. Otherwise the analysis is as shown in the following examples.

Example 7.9

Obtain an expression for the voltage at node 1 in the circuit of Fig. 7.3.

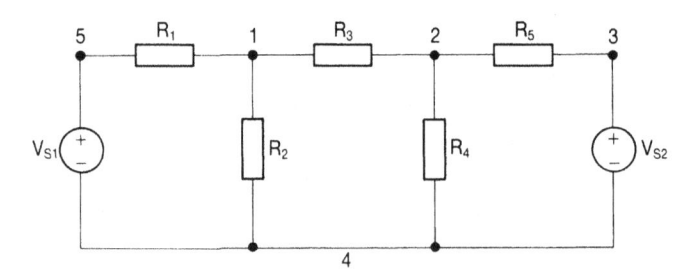

Figure 7.3

Solution

The nodes are identified as 1, 2, 3, 4 and 5, their voltages being V_1, V_2, V_3, V_4 and V_5, respectively. Node 4 is taken to be the reference, so $V_4 = 0$. Also we see that $V_5 = V_{S1}$ and that $V_3 = V_{S2}$. We therefore have two unknown node voltages, and to solve for them we need two equations.

Applying KCL to node 1 and using conductances rather than resistances we have, assuming all currents flow away from the node,

$$G_3(V_1 - V_2) + G_2(V_1 - V_4) + G_1(V_1 - V_5) = 0$$

$$G_3V_1 - G_3V_2 + G_2V_1 - 0 + G_1V_1 - G_1V_{S1} = 0$$

$$(G_1 + G_2 + G_3)V_1 - G_3V_2 = G_1V_{S1} \tag{7.5}$$

Applying KCL to node 2, again using conductances,

$$G_3(V_2 - V_1) + G_4(V_2 - V_4) + G_5(V_2 - V_3) = 0$$

$$G_3V_2 - G_3V_1 + G_4V_2 - 0 + G_5V_2 - G_5V_{S2} = 0$$

$$(G_3 + G_4 + G_5)V_2 - G_3V_1 = G_5V_{S2} \tag{7.6}$$

From Equations (7.5) and (7.6) we see a similar pattern to that in Equations (7.2) and (7.3) emerging: for node 1,

7.3 Nodal voltage analysis

- the coefficient of V_1 is the sum of all the conductances connected to node 1,
- the coefficient of V_2 is (-1) times the conductance connected between node 2 and node 1,
- the right-hand side of the equation is the current being fed directly into the node from source V_{S1};

and for node 2,

- the coefficient of V_2 is the sum of all the conductances connected to node 2,
- the coefficient of V_1 is (-1) times the conductance connected between node 1 and node 2,
- the right-hand side of the equation is the current being fed directly into the node from source V_{S2}.

In matrix form Equations (7.5) and (7.6) may be written

$$\begin{bmatrix} (G_1 + G_2 + G_3) & -G_3 \\ -G_3 & (G_3 + G_4 + G_5) \end{bmatrix} \begin{bmatrix} V_1 \\ V_2 \end{bmatrix} = \begin{bmatrix} V_{S1}G_1 \\ V_{S2}G_5 \end{bmatrix} \quad (7.7)$$

Using Cramer's rule to solve for V_1 we have $V_1 = \Delta_1/\Delta$. Now

$$\Delta = \begin{vmatrix} (G_1 + G_2 + G_3) & -G_3 \\ -G_3 & (G_3 + G_4 + G_5) \end{vmatrix}$$

The sum of the products of the elements along the diagonals to the right is

$$(G_1 + G_2 + G_3)(G_3 + G_4 + G_5)$$

The sum of the products of the elements along the diagonals to the left is $(-G_3)(-G_3) = G_3^2$, so

$$\Delta = (G_1 + G_2 + G_3)(G_3 + G_4 + G_5) - G_3^2$$

Now Δ_1 is Δ with the first column replaced by the column vector on the right-hand side of the Equation (7.7), so that

$$\Delta_1 = \begin{vmatrix} V_{S1}G_1 & -G_3 \\ V_{S2}G_5 & (G_3 + G_4 + G_5) \end{vmatrix}$$

The sum of the products of the elements along the diagonals to the right is $(V_{S1}G_1)(G_3 + G_4 + G_5)$. The sum of the products of the elements along the diagonals to the left is $(-G_3)(V_{S2}G_5)$. Therefore

$$\Delta_1 = (V_{S1}G_1)(G_3 + G_4 + G_5) - (-G_3)(V_{S2}G_5)$$
$$= (V_{S1}G_1)(G_3 + G_4 + G_5) + V_{S2}G_3G_5$$

Finally

$$V_1 = \Delta_1/\Delta$$
$$= [(V_{S1}G_1)(G_3 + G_4 + G_5) + (V_{S2}G_3G_5)]/(G_1 + G_2 + G_3)(G_3 + G_4 + G_5) - (G_3^2)]$$

Example 7.10

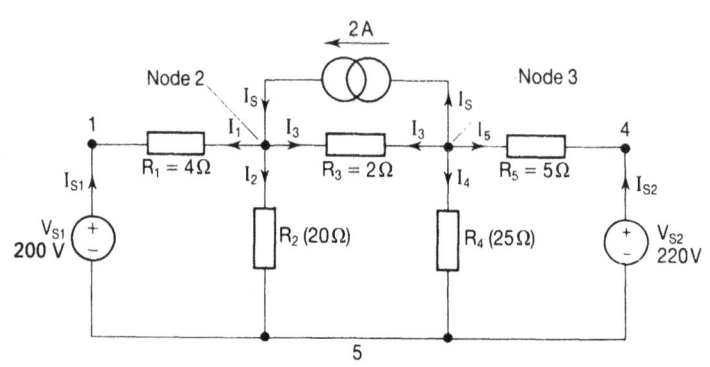

Figure 7.4

For the circuit of Fig. 7.4 determine (1) the potential difference across the resistor R_2, (2) the current supplied by the voltage source V_{S2}.

Solution

The nodes are identified as 1, 2, 3, 4 and 5, their respective voltages being V_1, V_2, V_3, V_4 and V_5. We choose node 5 to be the reference node and let $V_5 = 0$. Also

$$(V_1 - V_5) = (V_1 - 0) = V_1 = V_{S1} = 200 \text{ V}$$

and

$$(V_4 - V_5) = (V_4 - 0) = V_4 = V_{S2} = 220 \text{ V}$$

We therefore have two unknown voltages (V_2 and V_3) and we need two independent equations to solve for them.

Applying KCL to node 2 and assuming the currents to be flowing as shown, we have

$$I_1 + I_2 + I_3 - I_S = 0$$
$$(V_2 - V_1)/R_1 + (V_2 - V_5)/R_2 + (V_2 - V_3)/R_3 - I_S = 0$$

Using conductances we have

$$(V_2 - V_1)G_1 + G_2(V_2 - V_5) + G_3(V_2 - V_3) - I_S = 0$$
$$G_1V_2 - G_1V_{S1} + G_2V_2 - G_2V_5 + G_3V_2 - G_3V_3 = I_S$$
$$(G_1 + G_2 + G_3)V_2 - G_3V_3 = (I_S + G_1V_{S1})$$

Note that we could have written this equation down immediately without

7.3 Nodal voltage analysis 153

recourse to Kirchhoff's law by using the pattern noted in the bullet points leading to Equations (7.4) and (7.7) above. Thus:

- the coefficient of V_2 is the sum of all the conductances connected to node 2, i.e. $(G_1 + G_2 + G_3)$;
- the coefficient of V_3 is (-1) times the conductance connected between nodes 2 and 3, i.e. $(-G_3)$;
- the right-hand side of the equation is the current fed directly into the node from source V_{S1} $(G_1 V_{S1})$ and source I_S (I_S), i.e. a total of $(I_S + G_1 V_{S1})$.

Putting in the numbers we have

$$(1/4 + 1/20 + 1/2)V_2 - (1/2)V_3 = [2 + (1/4)200]$$

$$0.8 V_2 - 0.5 V_3 = 52 \tag{7.8}$$

For node 3:

- the coefficient of V_3 is the sum of all the conductances connected to node 3 $(G_3 + G_4 + G_5)$;
- the coefficient of V_2 is (-1) times the conductance connected between nodes 3 and 2 $(-G_3)$;
- the right-hand side of the equation is the total current fed directly into the node from the voltage source V_{S2} $(G_5 V_{S2})$ and the current source I_S $(-I_S)$ negative because the current is flowing *away* from the node.

Thus we have

$$(G_3 + G_4 + G_5)V_3 - G_3 V_2 = (G_5 V_{S2} - I_S)$$

Putting in the values we have

$$(1/2 + 1/25 + 1/5)V_3 - (1/2)V_2 = (1/5)220 - 2$$

$$0.74 V_3 - 0.5 V_2 = 42 \tag{7.9}$$

In matrix form Equations (7.8) and (7.9) become

$$\begin{bmatrix} 0.8 & -0.5 \\ -0.5 & 0.74 \end{bmatrix} \begin{bmatrix} V_2 \\ V_3 \end{bmatrix} = \begin{bmatrix} 52 \\ 42 \end{bmatrix} \tag{7.10}$$

Using Cramer's rule to solve for V_2,

$$V_2 = \Delta_1 / \Delta$$

$$\Delta = \begin{vmatrix} 0.8 & -0.5 \\ -0.5 & 0.74 \end{vmatrix}$$

$$= (0.8 \times 0.74) - (-0.5 \times -0.5) = 0.592 - 0.25 = 0.342$$

154 Nodal and mesh analysis

To determine Δ_1, we replace column 1 by the column vector on the right-hand side of Equation (7.10). Thus

$$\Delta_1 = \begin{vmatrix} 52 & -0.5 \\ 42 & 0.74 \end{vmatrix}$$

$= (52 \times 0.74) - (-0.5 \times 42) = 38.48 + 21 = 59.48$

Therefore

$V_2 = 59.48/0.342 = 173.9$ V

The potential difference across R_2 is

$V_2 - V_5 = 173.9 - 0 = 173.9$ V

Again using Cramer's rule, $V_3 = \Delta_2/\Delta$. Now to find Δ_2 we replace column 2 in Δ with the column vector on the right-hand side of Equation (7.10). Thus

$$\Delta_2 = \begin{vmatrix} 0.8 & 52 \\ -0.5 & 42 \end{vmatrix}$$

$= (0.8 \times 42) - (52 \times -0.5) = 33.6 + 26 = 59.6$

Therefore

$V_3 = 59.6/0.342 = 174.3$ V

The current supplied by the voltage source V_{S2} is given by

$I_{S2} = (V_4 - V_3)/R_5$
$= (V_{S2} - V_3)/R_5$
$= (220 - 174.3)/5$
$= 9.14$ A

Application to reactive a.c. circuits

The examples shown so far have been of purely resistive circuits. For a.c. circuits containing reactance the method applies equally but we must use complex impedances.

Example 7.11

Determine the potential differences across the admittance Y_2 in the circuit of Fig. 7.5. Shown overleaf.

Solution

The nodes are identified as 1, 2, 3, and 4, their voltages being V_1, V_2, V_3 and V_4. Let node 4 be the reference node so that $V_4 = 0$. Also

7.3 Nodal voltage analysis

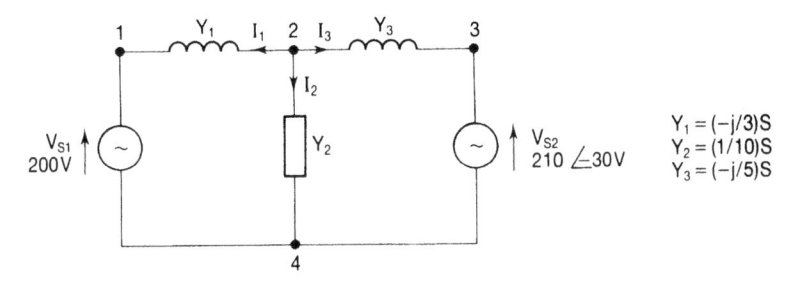

Figure 7.5

$Y_1 = (-j/3)$S
$Y_2 = (1/10)$S
$Y_3 = (-j/5)$S

$$V_1 - V_4 = V_1 = V_{S1} = 200\angle 0° \text{ V}$$

and

$$V_3 - V_4 = V_3 = V_{S2} = 210\angle -30° \text{ V}$$

We therefore need one equation in order to determine the one unknown voltage, V_2. Applying KCL to node 2 we have

$$I_1 + I_2 + I_3 = 0$$

$$Y_1(V_2 - V_1) + Y_2(V_2 - V_4) + Y_3(V_2 - V_3) = 0$$

$$Y_1 V_2 - Y_1 V_{S1} + Y_2 V_2 + Y_3 V_2 - Y_3 V_{S2} = 0$$

$$(Y_1 + Y_2 + Y_3)V_2 = Y_1 V_{S1} + Y_3 V_{S2}$$

$$V_2 = (Y_1 V_{S1} + Y_3 V_{S2})/(Y_1 + Y_2 + Y_3)$$

Now

$$Y_1 = 1/j3 = 1/(3\angle 90°) = 0.33\angle -90° \text{ S}$$

Therefore

$$Y_1 V_{S1} = 0.33\angle -90° \times 200\angle 0° = 66.7\angle -90° \text{ A} = (0 - j66.7) \text{ A}$$

Also

$$Y_3 = 1/j5 = 1/5\angle 90° = 0.2\angle -90° \text{ S}$$

Therefore

$$Y_3 V_{S2} = 0.2\angle -90° \times 210\angle -30° = 42\angle -120° = (-21 - j36.3) \text{ A}$$

Thus

$$Y_1 V_{S1} + Y_3 V_{S2} = (-21 - j103) \text{ A} = 105.1\angle -101.5° \text{ A}$$

Also we have

$$Y_1 + Y_2 + Y_3 = (1/j3) + (1/10) + (1/j5) = -j0.33 + 0.1 - j0.2$$
$$= (0.1 - j0.53) \text{ S} = 0.539\angle -79.3° \text{ S}$$

Thus

$V_2 = 105\angle -101.5°/0.539\angle -79.3° = 194.8\angle -22.2°$

The potential difference across the admittance Y_2 is

$V_2 - V_4 = V_2 = 194.8\angle -22.2°$ V.

Supernodes

If the voltage source V_S is connected between a node N and the reference node, the voltage of node N becomes V_S as we have seen in the previous two examples. If, however, the voltage source is connected between two nodes neither of which is the reference, we introduce the notion of a supernode.

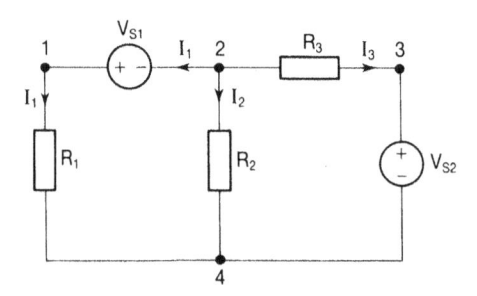

Figure 7.6

In the circuit of Fig. 7.6 the nodes 1 and 2 and the voltage source V_{S1} together form a supernode. Two other nodes are identified as 3 and 4. We choose node 4 as the reference and its voltage is $V_4 = 0$. Applying KCL to the supernode we have

$I_1 + I_2 + I_3 = 0$

$(V_1 - V_4)/R_1 + (V_2 - V_4)/R_2 + (V_2 - V_3)/R_3 = 0$

Now $(V_3 - V_4) = V_3 = V_{S2}$ so that, using conductances, we have

$G_1 V_1 + G_2 V_2 + G_3(V_2 - V_{S2}) = 0$

But $V_1 = V_{S1} + V_2$, therefore

$G_1(V_{S1} + V_2) + G_2 V_2 + G_3 V_2 - G_3 V_{S2} = 0$

$G_1 V_{S1} + G_1 V_2 + G_2 V_2 + G_3 V_2 - G_3 V_{S2} = 0$

$(G_1 + G_2 + G_3) V_2 = G_3 V_{S2} - G_1 V_{S1}$

$V_2 = (G_3 V_{S2} - G_1 V_{S1})/(G_1 + G_2 + G_3)$ \hfill (7.11)

Example 7.12

Determine the voltage across the resistors R_2 and R_4 in the circuit of Fig. 7.7.

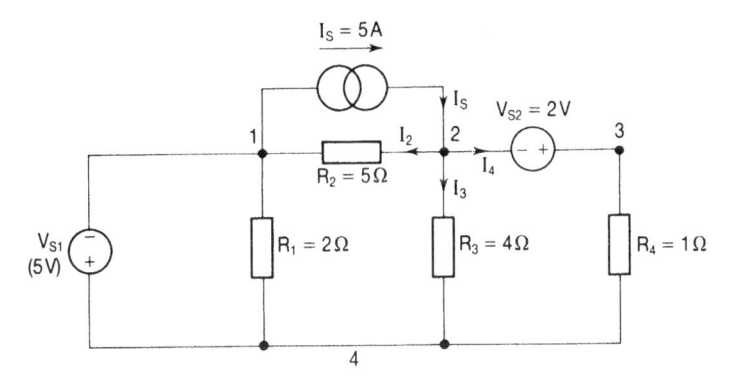

Figure 7.7

Solution

The voltages of the four nodes are V_1, V_2, V_3 and V_4. The reference node is node 4 and its voltage is $V_4 = 0$. Also

$$V_1 - V_4 = V_1 = V_{S1} = -5 \text{ V}$$

Nodes 2 and 3, together with the voltage source V_{S2}, constitute a supernode.
Applying KCL to the supernode we have

$$I_2 + I_3 + I_4 = I_S$$
$$(V_2 - V_1)/R_2 + (V_3 - V_4)/R_4 + (V_2 - V_4)/R_3 = I_S$$

Using conductances and with $V_4 = 0$ and $V_1 = V_{S1}$ we have

$$G_2 V_2 - G_2 V_{S1} + G_4 V_3 + G_3 V_2 = I_S$$
$$(G_2 + G_3)V_2 + G_4 V_3 = I_S + G_2 V_{S1}$$

But $V_3 = V_{S2} + V_2$, so

$$(G_2 + G_3)V_2 + G_4(V_{S2} + V_2) = I_S + G_2 V_{S1}$$
$$(G_2 + G_3 + G_4)V_2 = I_S + G_2 V_{S1} - G_4 V_{S2}$$
$$V_2 = (I_S + G_2 V_{S1} - G_4 V_{S2})/(G_2 + G_3 + G_4)$$

Putting in the values,

$I_S = 5$ A, $G_2 = (1/5)$ S, $G_3 = (1/4)$ S, $G_4 = (1/1)$ S, $V_{S1} = -5$ V and $V_{S2} = 2$ V

$$V_2 = (5 - 1 - 2)/(0.2 + 0.25 + 1) = 2/1.45 = 1.38 \text{ V}$$

The potential difference across the resistor R_2 is

$$V_2 - V_1 = 1.38 - (-5) = 6.38 \text{ V}.$$

The potential difference across the resistor R_4 is

$$V_3 - V_4 = (V_{S2} + V_2) - V_4 = 2 + 1.38 - 0 = 3.38 \text{ V}$$

7.4 MESH CURRENT ANALYSIS

Whereas in the nodal voltage method of analysis we used Kirchhoff's current law to set up equations from which we could determine the voltages at the various nodes, in the mesh current method of analysis we use Kirchhoff's voltage law to set up equations from which the currents in the various meshes can be calculated. To illustrate the method we will consider the two-mesh circuit of Fig. 7.8. Remember from Chapter 3 that meshes cannot have loops inside them so the loop containing V_{S1}, R_1, R_3 and V_{S2} is not a mesh. We assign the mesh currents I_1 and I_2 to the meshes 1 and 2. Note that the branch currents I_4 and I_5 are, respectively, equal to the mesh currents I_1 and I_2, while the branch current I_3 is $I_1 - I_2$.

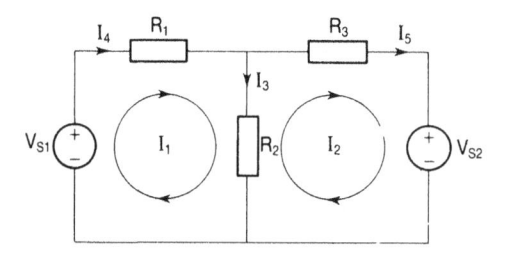

Figure 7.8

Applying KVL to mesh 1 and taking the clockwise direction to be positive, we have

$$V_{S1} - R_1 I_4 - R_2 I_3 = 0$$
$$V_{S1} - R_1 I_1 - R_2(I_1 - I_2) = 0$$
$$(R_1 + R_2)I_1 - R_2 I_2 = V_{S1} \tag{7.12}$$

Applying KVL to mesh 2 and taking the clockwise direction to be positive, we have

$$R_2 I_3 - R_3 I_5 - V_{S2} = 0$$
$$R_2(I_1 - I_2) - R_3 I_2 - V_{S2} = 0$$
$$R_2 I_1 - R_2 I_2 - R_3 I_2 - V_{S2} = 0$$
$$-R_2 I_1 + (R_2 + R_3)I_2 = -V_{S2} \tag{7.13}$$

In matrix form Equations (7.12) and (7.13) may be written

$$\begin{bmatrix} (R_1 + R_2) & -R_2 \\ -R_2 & (R_2 + R_3) \end{bmatrix} \begin{bmatrix} I_1 \\ I_2 \end{bmatrix} = \begin{bmatrix} V_{S1} \\ -V_{S2} \end{bmatrix} \tag{7.14}$$

Equations (7.12) and (7.13) can be solved simultaneously to determine I_1 and I_2. Alternatively, Cramer's rule can be applied to Equation (7.14).

Let us now consider the circuit of Fig. 7.9. This is a three-mesh circuit, the

7.4 Mesh current analysis

mesh currents being I_1, I_2 and I_3. None of the other loops are meshes because they have other loops inside them. We note that the branch currents I_6, I_7 and I_8 are the mesh currents I_1, I_2 and I_3, respectively. Also the branch current I_4 is the difference of two mesh currents $(I_1 - I_2)$. Similarly the branch current I_5 is the difference of two mesh currents $(I_2 - I_3)$.

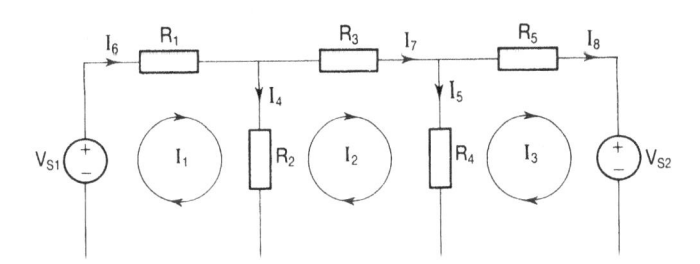

Figure 7.9

Applying KVL to mesh 1:

$$V_{S1} - R_1 I_1 - R_2 I_4 = 0$$
$$V_{S1} - R_1 I_1 - R_2(I_1 - I_2) = 0$$
$$V_{S1} - R_1 I_1 - R_2 I_1 + R_2 I_2 = 0$$

$$(R_1 + R_2) I_1 - R_2 I_2 = V_{S1} \quad (7.15)$$

Applying KVL to mesh 2:

$$R_2 I_4 - R_3 I_2 - R_4 I_5 = 0$$
$$R_2(I_1 - I_2) - R_3 I_2 - R_4(I_2 - I_3) = 0$$
$$R_2 I_1 - R_2 I_2 - R_3 I_2 - R_4 I_2 + R_4 I_3 = 0$$

$$-R_2 I_1 + (R_2 + R_3 + R_4) I_2 - R_4 I_3 = 0 \quad (7.16)$$

Applying KVL to mesh 3:

$$R_4 I_5 - R_5 I_3 - V_{S2} = 0$$
$$R_4(I_2 - I_3) - R_5 I_3 - V_{S2} = 0$$
$$R_4 I_2 - (R_4 + R_5) I_3 = V_{S2}$$

Multiplying throughout by -1 we have

$$-R_4 I_2 + (R_4 + R_5) I_3 = -V_{S2} \quad (7.17)$$

In matrix form Equations (7.15), (7.16) and (7.17) may be written

$$\begin{bmatrix} (R_1 + R_2) & -R_2 & 0 \\ -R_2 & (R_2 + R_3 + R_4) & -R_4 \\ 0 & -R_4 & (R_4 + R_5) \end{bmatrix} \begin{bmatrix} I_1 \\ I_2 \\ I_3 \end{bmatrix} = \begin{bmatrix} V_{S1} \\ 0 \\ -V_{S2} \end{bmatrix} \quad (7.18)$$

Equations (7.15), (7.16) and (7.17) can be solved simultaneously to determine

160 Nodal and mesh analysis

the three mesh currents or Cramer's rule can be used to solve the matrix equation (7.18).

We note from Equations (7.15), (7.16) and (7.17) that to form a particular mesh equation we:

1. multiply that mesh current by the sum of all the resistances around that mesh;

2. subtract the product of each adjacent mesh current and the resistance common to both meshes;

3. equate this to the voltage in the mesh, the sign being positive if the voltage source acts in the same direction as the mesh current and negative otherwise.

Example 7.13

For the circuit of Fig. 7.10 write down the three mesh equations from which the mesh currents I_1, I_2 and I_3 could be determined.

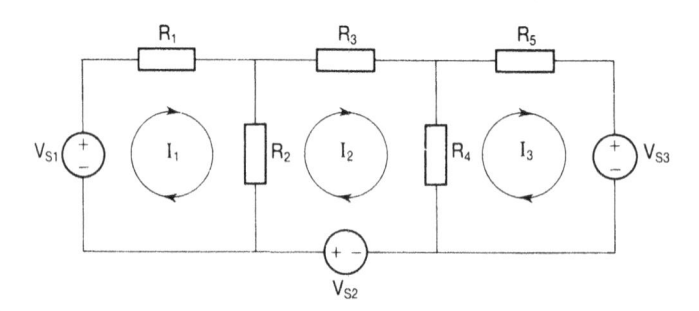

Figure 7.10

Solution

To set up the mesh equations we follow the three steps outlined above.
For mesh 1

1. The coefficient of I_1 is the sum of the resistances around the mesh ($=R_1 + R_2$). We therefore have $(R_1 + R_2)I_1$ on the left-hand side of the equation.

2. There is one adjacent mesh whose current is I_2. The coefficient of I_2 is minus the resistance common to meshes 1 and 2 (i.e. $-R_2$). We thus have $-R_2 I_2$ on the left-hand side.

3. The right-hand side of the equation is V_{S1}, positive, as the source acts in the same direction as the mesh current.

The mesh equation is therefore

$$(R_1 + R_2)I_1 - R_2 I_2 = V_{S1} \tag{7.19}$$

7.4 Mesh current analysis

For mesh 2

1. The coefficient of the mesh current is the sum of the resistances around the mesh ($=R_2 + R_3 + R_4$). We therefore have $(R_2 + R_3 + R_4)I_2$ on the left-hand side of the equation.

2. There are two adjacent meshes (1 and 3). The resistance common to meshes 2 and 1 is R_2 and the resistance common to meshes 2 and 3 is R_4. We therefore have terms $-R_2I_1$ and $-R_4I_3$ on the left-hand side.

3. There is a voltage source V_{S2} acting in the same direction as the mesh current so $+V_{S2}$ appears on the right-hand side of the equation.

The mesh equation is therefore

$$-R_2I_1 + (R_2 + R_3 + R_3)I_2 - R_4I_3 = V_{S2} \tag{7.20}$$

For mesh 3

1. The total resistance around the mesh is $(R_4 + R_5)$ so we have $(R_4 + R_5)I_3$ on the left-hand side of the equation.

2. There is one adjacent mesh (2) and the resistance common to it and mesh 3 is R_4. We therefore have a term $-R_4I_2$ appearing on the left-hand side.

3. The voltage source V_{S3} acts in the opposite direction to the mesh current so that $-V_{S3}$ appears on the right-hand side.

The mesh equation is therefore

$$-R_4I_2 + (R_4 + R_5)I_3 = -V_{S3} \tag{7.21}$$

In matrix form the equations may be written

$$\begin{bmatrix} (R_1 + R_2) & -R_2 & 0 \\ -R_2 & (R_2 + R_3 + R_4) & -R_4 \\ 0 & -R_4 & (R_4 + R_5) \end{bmatrix} \begin{bmatrix} I_1 \\ I_2 \\ I_3 \end{bmatrix} = \begin{bmatrix} V_{S1} \\ V_{S2} \\ -V_{S3} \end{bmatrix} \tag{7.22}$$

Example 7.14

Determine the currents supplied by the voltage sources V_{S1} and V_{S2} in the circuit of Fig. 7.11.

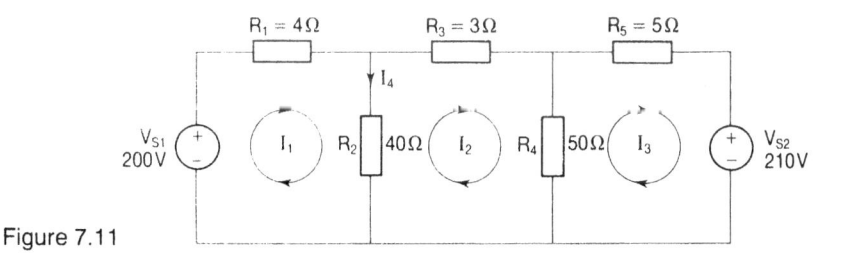

Figure 7.11

Solution

There are three meshes identified by the currents I_1, I_2 and I_3. Applying KVL to mesh 1 and taking the clockwise direction to be positive, we have

$$V_{S1} - R_1 I_1 - R_2 I_4 = 0$$

The branch current I_4 is $I_1 - I_2$, so

$$V_{S1} - R_1 I_1 - R_2(I_1 - I_2) = 0$$
$$V_{S1} - R_1 I_1 - R_2 I_1 + R_2 I_2 = 0$$
$$(R_1 + R_2)I_1 - R_2 I_2 = V_{S1} \tag{7.23}$$

Note that we could have written down this equation immediately using the three steps outlined above.

For mesh 2, which has two adjacent meshes, steps (1) and (2) give $(R_2 + R_3 + R_4)I_2 - R_2 I_1 - R_4 I_3$ for the left-hand side of the equation. There is no voltage source in this mesh so the right-hand side is simply zero. The mesh equation is therefore

$$-R_2 I_1 + (R_2 + R_3 + R_4)I_2 - R_4 I_3 = 0 \tag{7.24}$$

For mesh 3, which has only one adjacent mesh, the left-hand side of the equation is

$$-R_4 I_2 + (R_4 + R_5)I_3$$

The voltage source is acting in the opposite direction to the mesh current so the right-hand side of the equation is $-V_{S2}$. The mesh equation is therefore

$$-R_4 I_2 + (R_4 + R_5)I_3 = -V_{S2} \tag{7.25}$$

Putting in the values for the resistances and voltages. Equations (7.23), (7.24) and (7.25) may be written in matrix form as

$$\begin{bmatrix} (4+40) & -40 & 0 \\ -40 & (40+3+50) & -50 \\ 0 & -50 & (50+5) \end{bmatrix} \begin{bmatrix} I_1 \\ I_2 \\ I_3 \end{bmatrix} = \begin{bmatrix} 200 \\ 0 \\ -210 \end{bmatrix} \tag{7.26}$$

Using Cramer's rule $I_1 = \Delta_1/\Delta$. Now

$$\Delta = \begin{vmatrix} 44 & -40 & 0 \\ -40 & 93 & -50 \\ 0 & -50 & 55 \end{vmatrix}$$

$$= [(44 \times 93 \times 55) + (-40 \times -50 \times 0) + (0 \times -40 \times -50)]$$
$$- [(0 \times 93 \times 0) + (-40 \times -40 \times 55) + (44 \times -50 \times -50)]$$
$$= 225\,060 - (88\,000 + 110\,000)$$
$$= 27\,060$$

7.4 Mesh current analysis

To find Δ_1 we replace column 1 of Δ by the column vector on the right-hand side of the matrix equation. Thus

$$\Delta_1 = \begin{vmatrix} 200 & -40 & 0 \\ 0 & 93 & -50 \\ -210 & -50 & 55 \end{vmatrix}$$

$= [(200 \times 93 \times 55) + (-40 \times -50 \times -210) + (0 \times 0 \times -50)]$
$\quad - [(0 \times 93 \times -210) + (-40 \times 0 \times 55) + (200 \times -50 \times -50)]$
$= (603\,000) - (500\,000)$
$= 103\,000$
$I_1 = 103\,000/27\,060 = 3.81$ A.

To find Δ_3 we replace the third column of Δ by the column vector giving

$$\Delta_3 = \begin{vmatrix} 44 & -40 & 200 \\ -40 & 93 & 0 \\ 0 & -50 & -210 \end{vmatrix}$$

$= [(44 \times 93 \times -210) + (-40 \times -0 \times 0) + (200 \times -40 \times -50)]$
$\quad - [(200 \times 93 \times 0) + (-40 \times -40 \times -210) + (44 \times 0 \times -50)]$
$= (-859\,320 + 400\,000) - (-336\,000)$
$= -123\,320$
$I_3 = \Delta_3/\Delta = -123\,320/27\,060 = -4.56$ A

The minus sign indicates that the current is flowing in the opposite direction to that shown in the circuit diagram (i.e. it is flowing out of the positive terminal of the voltage source).

Circuits containing voltage and current sources

If the current source is located in one mesh only, then that mesh current is the source current and the analysis is quite straightforward, as shown by the following example.

Example 7.15

Determine the current through the resistor R_2 in the circuit of Fig. 7.12.

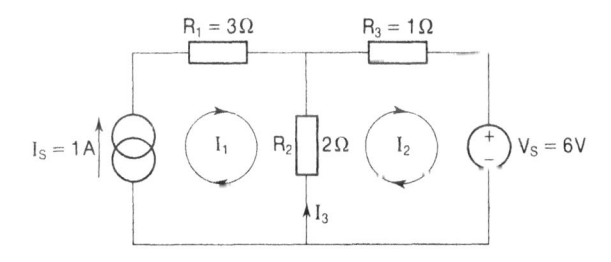

Figure 7.12

Solution

There are two meshes whose currents are I_1 and I_2, respectively, and we note that the mesh current I_1 is equal to the current source I_S. Applying KVL to mesh 2 and taking the anticlockwise direction to be positive, we have

$$V_S + R_3I_2 + R_2I_3 = 0$$

Since $I_3 = I_2 - I_1$ then

$$V_S + R_3I_2 + R_2I_2 - R_2I_1 = 0$$
$$(R_2 + R_3)I_2 = R_2I_1 - V_S$$
$$= R_2I_s - V_S$$
$$I_2 = (R_2I_s - V_s)/(R_2 + R_3)$$

Putting in the values we have

$$I_2 = [(2 \times 1) - 6]/(2 + 1) = (-4/3) \text{ A}$$

The current through the resistor R_2 is

$$I_3 = I_2 - I_1 = I_2 - I_S = (-4/3) - 1 = -7/3 = -2.33 \text{ A}$$

The negative sign indicates that the current is flowing in the opposite direction to that shown in the circuit diagram (i.e. it flows downwards through the resistor).

Supermeshes

We saw that in nodal voltage analysis two nodes having a voltage source connected between them constitute a supernode. In mesh current analysis two meshes which have a current source common to both of them form what is known as a supermesh. As an example, consider the circuit of Fig. 7.13 which has three meshes identified by the currents I_1, I_2 and I_3. Meshes 2 and 3 have a common current source I_S and together they constitute a supermesh.

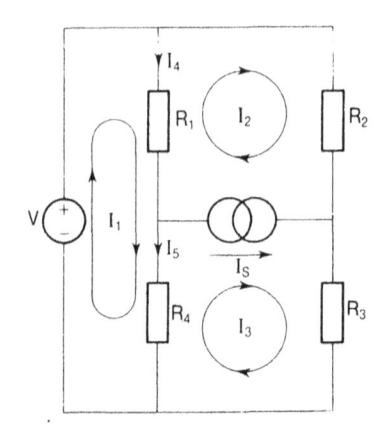

Figure 7.13

Applying KVL to mesh 1 and taking the clockwise direction to be positive, we have

$$V - R_1 I_4 - R_4 I_5 = 0$$

The branch current I_4 is $I_1 - I_2$ and the branch current I_5 is $I_1 - I_3$. Therefore

$$V - R_1(I_1 - I_2) - R_4(I_1 - I_3) = 0$$
$$V - R_1 I_1 + R_1 I_2 - R_4 I_1 + R_4 I_3 = 0$$

Rearranging,

$$(R_1 + R_4)I_1 - R_1 I_2 - R_4 I_3 = V \qquad (7.27)$$

Note again that:

1. the mesh current is multiplied by the sum of the resistances around the mesh;

2. there are two adjacent meshes and for each of these we subtract the product of the mesh current and the resistance common to it and mesh 1;

3. the right-hand side of the equation is the voltage source in the mesh and is positive because it acts in the same direction as the mesh current.

Applying KVL to the supermesh and taking the clockwise direction to be positive, we have

$$R_1 I_4 - R_2 I_2 - R_3 I_3 + R_4 I_5 = 0$$
$$R_1(I_1 - I_2) - R_2 I_2 - R_3 I_3 + R_4(I_1 - I_3) = 0$$
$$R_1 I_1 - R_1 I_2 - R_2 I_2 - R_3 I_3 + R_4 I_1 - R_4 I_3 = 0$$
$$(R_1 + R_4)I_1 - (R_1 + R_2)I_2 - (R_3 + R_4)I_3 = 0$$

Multiplying throughout by -1 and rearranging,

$$(R_1 + R_2)I_2 + (R_3 + R_4)I_3 - (R_1 + R_4)I_1 = 0 \qquad (7.28)$$

Note that in this case:

1. the supermesh current I_2 is multiplied by the total resistance through which it flows and the supermesh current I_3 is multiplied by the total resistance through which *it* flows and these products are added;

2. there is one adjacent mesh and its current is multiplied by the total resistance through which it flows and this product is subtracted;

3. there is no voltage source in the supermesh, so that the right-hand side of the equation is zero.

The 'three-step' method could thus have been used to write down Equations (7.27) and (7.28) immediately without recourse to Kirchhoff's voltage law. We see that the current in the branch common to meshes 2 and 3 is I_5 ($= I_3 - I_2$), so

166 Nodal and mesh analysis

$$I_3 = I_S + I_2 \tag{7.29}$$

Substituting for I_3 from Equation (7.29) in Equation (7.27) we have

$$(R_1 + R_4)I_1 - R_1I_2 - R_4(I_S + I_2) = V$$
$$(R_1 + R_4)I_1 - R_1I_2 - R_4I_S - R_4I_2 = V$$

$$(R_1 + R_4)I_1 - (R_1 + R_4)I_2 = V + R_4I_S \tag{7.30}$$

Substituting for I_3 from Equation (7.29) in Equation (7.28) we have

$$(R_1 + R_2)I_2 + (R_3 + R_4)(I_S + I_2) - (R_1 + R_4)I_1 = 0$$
$$(R_1 + R_2)I_2 + R_3I_S + R_3I_2 + R_4I_S + R_4I_2 - (R_1 + R_4)I_1 = 0$$

$$(R_1 + R_2 + R_3 + R_4)I_2 - (R_1 + R_4)I_1 = -(R_3 + R_4)I_S \tag{7.31}$$

We now have two equations (7.30) and (7.31) from which I_1 and I_2 can be determined. Once I_2 is known I_3 follows immediately from Equation (7.29).

Example 7.16

Determine (1) the mesh current I_2 and (2) the voltage at node X in the circuit of Fig. 7.14.

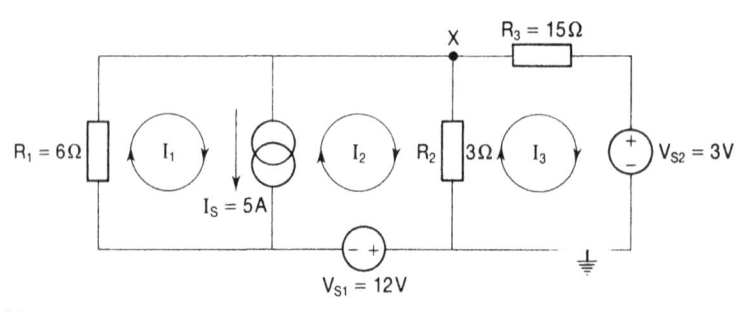

Figure 7.14

Solution

There are three meshes in total but since there is a current source common to meshes 1 and 2, these can be considered to be a supermesh. Applying the three-step method to the supermesh:

1. we multiply the mesh current I_1 by the resistance through which it flows (R_1) and add this to the product of the mesh current I_2 and the resistance through which *it* flows (R_2);

2. there is one adjacent mesh whose current is I_3. We multiply this by the resistance (R_2) common to the two meshes and subtract the product;

3. there is a voltage source (V_{S1}) which acts in the opposite direction to the mesh current flowing through it so the right-hand side of the supermesh equation is $-V_{S1}$

The supermesh equation is therefore

$$R_1I_1 + R_2I_2 - R_2I_3 = -V_{S1}$$

Now $I_1 - I_2 = I_S \Rightarrow I_1 = I_S + I_2$, so

$$R_1(I_S + I_2) + R_2I_2 - R_2I_3 = -V_{S1}$$

Putting in the values,

$$6(5 + I_2) + 3I_2 - 3I_3 = -12$$
$$9I_2 - 3I_3 = -42$$

Dividing through by 3 we get

$$3I_2 - I_3 = -14 \tag{7.32}$$

For mesh 3:

1. we multiply the mesh current by the resistance through which it flows $(R_2 + R_3)$;
2. there is one adjacent mesh whose current (I_2) is multiplied by the resistance common to both meshes (R_2);
3. the voltage source V_{S2} acts in the opposite direction to the current in the mesh so the right-hand side of the mesh equation is $-V_{S2}$.

The mesh equation is therefore

$$(R_2 + R_3)I_3 - R_2I_2 = -V_{S2}$$

Putting in the numbers and rearranging we have

$$-3I_2 + 18I_3 = -3 \tag{7.33}$$

Adding Equations (7.32) and (7.33) we have

$$17I_3 = -17 \quad \text{and} \quad I_3 = -1 \text{ A}$$

The minus sign indicates that the mesh current I_3 flows in the opposite direction to that shown in the diagram (i.e. anticlockwise rather than clockwise).

(1) To find the mesh current I_2

Substituting for $I_3 = -1$ A in Equation (7.32) we have $3I_2 - (-1) = -14$ and

$$I_2 = -15/3 = -5 \text{ A}$$

Again the minus sign indicates that the mesh current I_2 flows in an anticlockwise direction around the mesh.

(2) To find the voltage at node X

The voltage at node X is given by

$$V_X = V_{S2} + I_3R_3 = 3 + (-1) \times 15 = 3 - 15 = -12 \text{ V}$$

7.5 SELF-ASSESSMENT TEST

1. Define a matrix.
2. Explain what is meant by a column matrix.
3. Write down the row matrix having elements 1, 7, 9, and 81.
4. Express in double subscript form the element located in the third row and the fourth column of a matrix A.
5. For the matrices A and B state whether $A + B = B + A$.
6. For the matrices A and B state whether $AB = BA$.
7. Define the minor of an element of a matrix.
8. Define the cofactor of an element of a matrix.
9. Define the determinant of a matrix.
10. State Cramer's rule.
11. Outline the method of nodal voltage analysis of electric circuits.
12. What determines the choice of the reference node in nodal voltage analysis?
13. Define a supernode.
14. Outline the method of mesh current analysis of electric circuits.
15. Define a supermesh.

7.6 PROBLEMS

1. For matrix $A = \begin{bmatrix} 1 & 3 \\ 5 & 6 \end{bmatrix}$ and matrix $B = \begin{bmatrix} 1 & 0 \\ -1 & -2 \end{bmatrix}$ determine (a) $A + B$, (b) $A - B$, (c) AB, (d) BA.

2. Write out the following set of linear equations in the matrix form $AX = B$:
   (a) $2x - 3y = 1$
   $x + 5y = 7$
   (b) $x + y + 3z = 7$
   $x - 2y + z = 6$
   $-4y - 3z = 4.$

3. Find the linear equations represented by the following matrix equations
   (a) $\begin{bmatrix} 2 & 5 \\ 1 & 3 \end{bmatrix} \begin{bmatrix} x \\ y \end{bmatrix} = \begin{bmatrix} 27 \\ 16 \end{bmatrix}$

(b) $\begin{bmatrix} 1 & -1 & -5 \\ 1 & 1 & 1 \\ -1 & 0 & -3 \end{bmatrix} \begin{bmatrix} x \\ y \\ z \end{bmatrix} = \begin{bmatrix} 6 \\ 0 \\ 1 \end{bmatrix}$

4 Find the determinant of the matrix $\begin{bmatrix} 3 & 5 \\ 4 & 8 \end{bmatrix}$

5 Evaluate the determinant of the matrix

$A = \begin{bmatrix} 2 & 7 & 6 \\ 1 & 3 & 4 \\ 5 & 8 & 9 \end{bmatrix}$

6 Write down and evaluate the minor of the element a_{21} of the matrix A in Problem 5.

7 Write down and evaluate the cofactor of the element a_{32} of the matrix A in Problem 5.

NB: Each of the following problems should be solved using (a) nodal voltage analysis and (b) mesh current analysis even though in some cases other methods might be more appropriate.

8 Determine the value of the current I through the 1 Ω resistor in the circuit of Fig. 7.15.

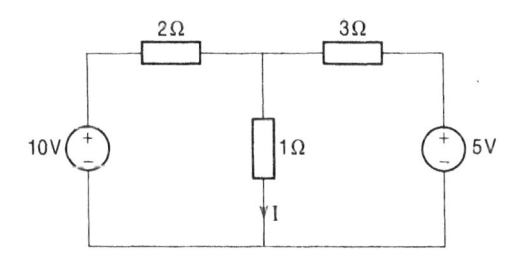

Figure 7.15

9 Calculate the value of the current flowing in the resistor R (= 30 Ω) in the circuit of Fig. 7.16

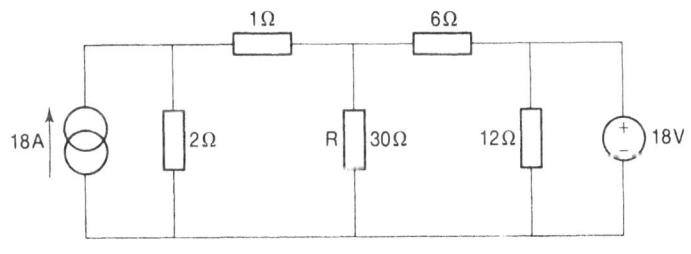

Figure 7.16

170 Nodal and mesh analysis

10 Obtain the value of the current flowing through resistor R ($= 1\,\Omega$) in the circuit of Fig. 7.17

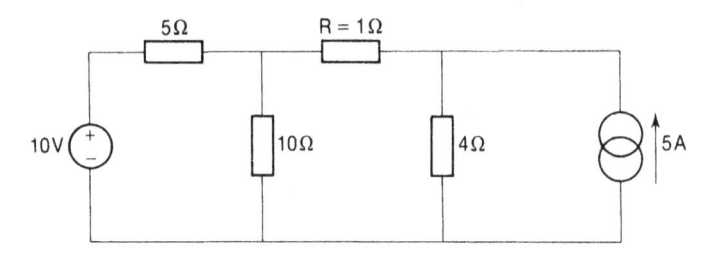

Figure 7.17

11 Find the value of the current I flowing through the 40 Ω resistor in the circuit of Fig. 7.18.

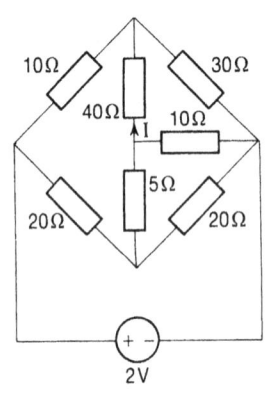

Figure 7.18

12 Calculate (a) the loop currents I_1, I_2 and I_3 and (b) the total power consumed in the circuit of Fig. 7.19.

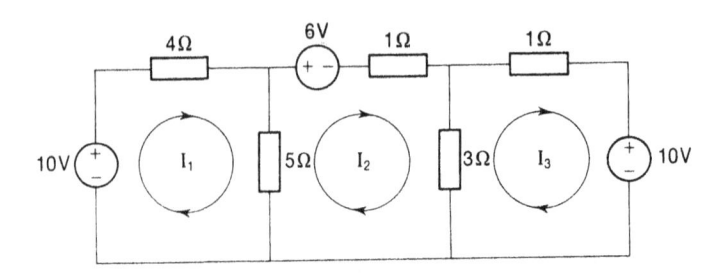

Figure 7.19

13 Determine the current *I* supplied by the voltage source in the circuit of Fig. 7.20.

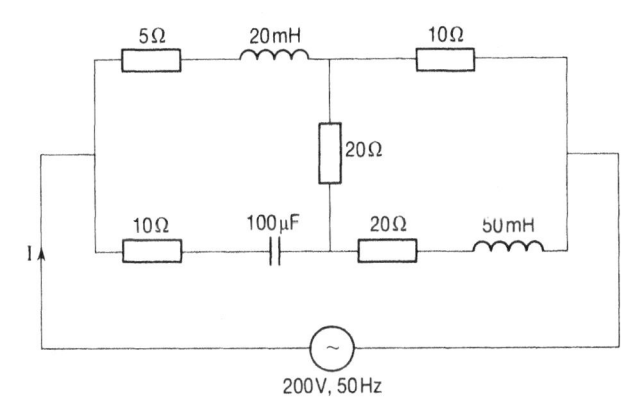

Figure 7.20

8 Transient analysis

8.1 INTRODUCTION

We saw in Chapter 2 that the potential difference across a capacitor cannot change instantaneously (Equation 2.26) and that the current through an inductor cannot change suddenly (Equation 2.35). If, therefore, when a circuit containing capacitance or/and inductance is operating in the steady state and conditions change for some reason, requiring the current and voltage values to change, there will be a finite period of time during which these changes take place. This period is called a period of transient operation.

Two obvious examples of transient operation are (1) when a circuit containing capacitance or inductance is initially switched on and (2) when such a circuit, having been operating in the steady state for some time, is suddenly switched off.

These transient conditions are associated with the changes in the energy stored in the capacitor or the inductor, and circuits containing either of these elements are called single energy circuits. Circuits containing both of these elements are called double energy circuits. Because there is no energy stored in purely resistive circuits, currents and voltages in such circuits are able to change without periods of transient operation (i.e. instantaneously).

8.2 CIRCUITS CONTAINING RESISTANCE AND INDUCTANCE

The sudden application of a step voltage

The series RL circuit shown overleaf in Fig. 8.1 is connected to a d.c. voltage source V such that when the switch S is closed (at an instant $t = 0$ say) a voltage V is suddenly applied to the circuit. This is called a step function and is shown in Fig. 8 2. For this step function,

$V(t) = 0$ for $t < 0$

$V(t) = V$ for $t \geq 0$

8.2 Circuits containing resistance and inductance

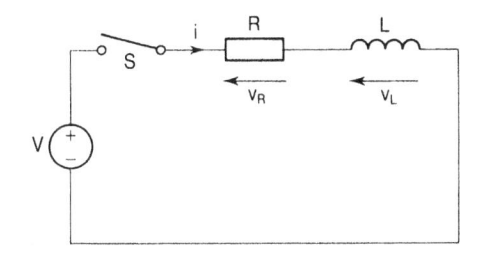

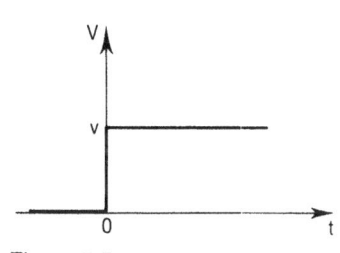

Figure 8.1

Figure 8.2

Applying KVL to the circuit we see that

$$V - V_R - V_L = 0$$

where

$$V_L = L(di/dt)$$

and

$$V_R = iR$$

so that

$$V - iR - L(di/dt) = 0 \tag{8.1}$$

Rearranging, we get

$$di/dt = (V - iR)/L = [(V/R) - i]/(L/R)$$

Separating the variables di and dt we have

$$di/[(V/R) - i] = R\,dt/L$$

Integrating, and remembering that $\int (dx/x) = \ln x + $ a constant, and that $\int dx = x + $ a constant, we have

$$-\ln[(V/R) - i] = (R/L)t + C$$

where C is a constant. Therefore

$$\ln[(V/R) - i] = -(R/L)t - C$$

Taking antilogs we get

$$(V/R) - i = \exp[(-R/L)t - C] = \exp(-Rt/L)\exp(-C)$$

Now at $t = 0$, $i = 0$ so that $\exp(-C) = V/R$. Therefore

$$(V/R) - i = \exp(-Rt/L) \times V/R$$
$$i = V/R - (V/R)\exp(-Rt/L)$$
$$= (V/R)[1 - \exp(-Rt/L)]$$

(V/R) is the value reached by the current when all transients have died away, i.e. it is the steady state value (I) of the current, so we have finally

$$i = I[1 - \exp(-Rt/L)] \tag{8.2}$$

The time constant of an RL circuit

From Equation (8.2) we note that when $t = L/R$ seconds,

$$i = I[1 - \exp(-1)] = 0.632\,I \tag{8.3}$$

L/R is called the time constant of the circuit. Its symbol is τ and it is measured in seconds:

$$\tau = L/R \tag{8.4}$$

From Equation (8.3) we see that after a time equal to the time constant following the sudden application of a step function, the current will have reached 63.2 per cent of its steady state value. After a time equal to five time constants (5τ) the current will have reached $[1 - \exp(-5)] = 0.993$ or 99.3 per cent of its steady state value. Since mathematically the current can never reach I we say that to all intents and purposes it has done so after 5τ.

The voltage across the resistor is $v_R = iR = I[1 - \exp(-Rt/L)]R$ and, since $IR = V$,

$$v_R = V[1 - \exp(-Rt/L)] \tag{8.5}$$

The voltage across the inductor is $v_L = V - v_R = V - V[1 - \exp(-Rt/L)]$, so

$$v_L = V \exp(-Rt/L) \tag{8.6}$$

Equation (8.2) describes an exponential growth of current, Equation (8.5) describes an exponential growth of voltage and Equation (8.6) describes an exponential decay of voltage. The graphs of these are shown in Figs 8.3 and 8.4.

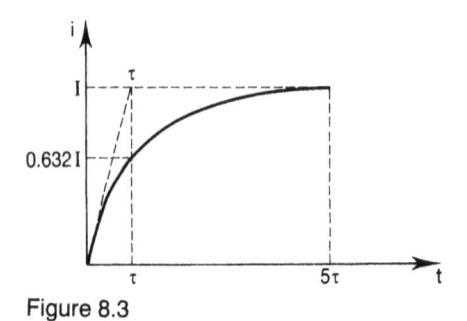

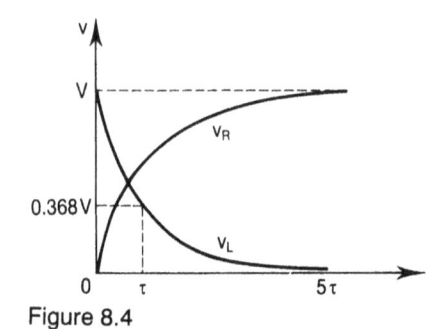

Figure 8.3

Figure 8.4

The rate of change of current is obtained by differentiating Equation (8.2). Thus, remembering that $I = V/R$

$$\begin{aligned}di/dt &= (V/R)(R/L)\exp(-Rt/L) \\ &= (V/L)\exp(-Rt/L)\end{aligned} \tag{8.7}$$

At $t = 0$, the rate of change of current is V/L amperes per second and if the current maintained this rate of growth for a time equal to the time constant of the circuit it would reach

$(V/L) \times \tau = (V/L) \times (L/R)$
$= V/R$ amperes

This is its final steady state value I.

Example 8.1

A d.c. voltage of 200 V is suddenly applied to a circuit consisting of a resistor of 20 Ω resistance in series with an inductor having an inductance of 0.5 H. Determine (1) the time constant of the circuit, (2) the final steady state value of the current, (3) the value of the current after a time equal to three time constants.

Solution

1 From Equation (8.4) the time constant is given by
   $L/R = 0.5/20 = 0.025$ s $= 25$ ms

2 The final (steady state) value of the current is given by
   $I = V/R = 200/20 = 10$ A

3 The time equivalent to 3τ is $3 \times 25 = 75$ ms. From Equation (8.2)
   $i = I[1 - \exp(-Rt/L)]$.
   With $I = 10$ A, $R = 20$ Ω, $L = 0.5$ H and $t = 0.075$ s, we obtain
   $i_{3\tau} = 10[1 - \exp(-20 \times 0.075/0.5)] = 9.5$ A

Example 8.2

A resistor having a resistance of 2 Ω is connected in series with an inductor of 20 H inductance. A step voltage is suddenly applied to the series combination and the initial rate of rise of current is 4 A/s. Determine (1) the time constant of the circuit, (2) the magnitude of the applied voltage.

Solution

The circuit is as shown in Figure 8.1 with $R = 2$ Ω and $L = 20$ H.

1 From Equation (8.4) the time constant is given by $\tau = L/R = 20/2 = 10$ s.

2 From Equation (8.7) with $t = 0$ we see that the initial rate of rise of current is given by V/L. It follows that $V =$ (initial rate of rise of current) $\times L$
   $= 4 \times 20 = 80$ V.

The sudden disconnection of a d.c. supply

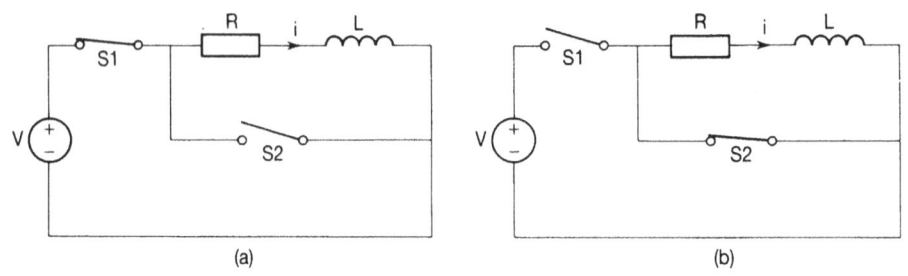

Figure 8.5

Suppose that the circuit of Fig. 8.5(a) has been in steady state operation for a long time ('long' here means long compared to the time constant of the circuit), when the switch S_1 is suddenly opened and S_2 is simultaneously closed. Let this instant be $t = 0$ and at this instant the current will have its steady state value (I). The circuit then becomes that of Fig. 8.5(b).

Applying KVL to the closed circuit we have

$$iR + L\,di/dt = 0$$
$$iR = -L\,di/dt$$

Separating the variables di and dt,

$$(R/L)\,dt = -(di/i)$$
$$\int (R/L)\,dt = -\int (1/i)\,di$$

Integrating gives

$$Rt/L = -\ln i + C \tag{8.8}$$

where C is a constant. Now at the instant $t = 0$, $i = I$ and so $0 = -\ln I + C$, so

$$C = \ln I$$

Substituting for C in Equation (8.8)

$$Rt/L = -\ln i + \ln I = \ln I/i$$

Taking antilogs, we have $\exp(Rt/L) = I/i$ so that $i = I/\exp(Rt/L)$. Finally

$$i = I \exp(-Rt/L) \tag{8.9}$$

The potential difference across the resistor is given by $v_R = iR = I \exp(-Rt/L)R$.
Since $IR = V$, then

$$v_R = V \exp(-Rt/L). \tag{8.10}$$

The voltage across the inductor is $v_L = -v_R = -V \exp(-Rt/L)$, so

$$v_L = -V \exp(-Rt/L) \tag{8.11}$$

8.2 Circuits containing resistance and inductance

Equation (8.9) represents an exponential decay of current, Equation (8.10) represents an exponential decay of voltage and Equation (8.11) describes an exponential rise of voltage starting from $-V$ and aiming towards zero. These are illustrated in Fig. 8.6(a) and (b).

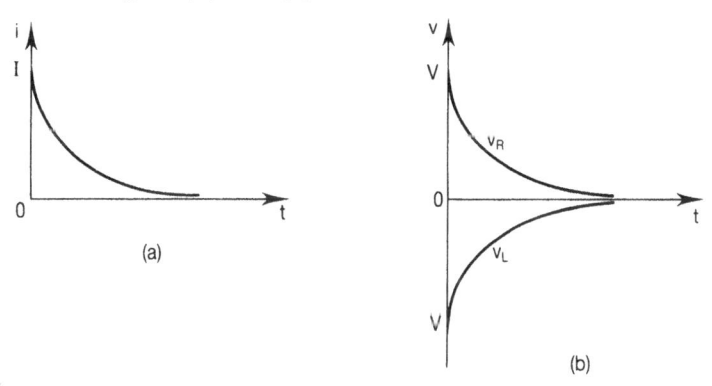

Figure 8.6

Example 8.3

A circuit consisting of a resistor having a resistance of 2 Ω in series with an inductor whose inductance is 10 H is fed from a 12 V d.c. supply. Thirty seconds after the circuit has been switched on a fault causes the supply to become short circuited. Determine the current in the circuit 2.5 s after the occurrence of the fault.

Solution

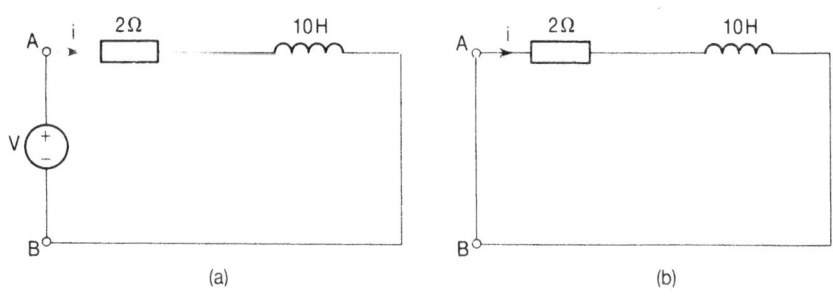

Figure 8.7

Fig. 8.7(a) shows the circuit before the fault occurs and Fig. 8.7(b) shows the circuit after the fault. The time constant of the circuit is given by $\tau = L/R = 10/2 = 5$ s. After 30 s (which is 6 τ) therefore, the current will have reached its steady state value:

$I = V/R = 12/2 = 6$ A

When the fault occurs $I = 6$ A and 2.5 s later

$i = I \exp(-Rt/L) = 6 \exp(-2 \times 2.5/10) = 3.64$ A

The response to the application of a pulse

Suppose that the pulse of width T seconds ($T \geq 5\tau$) shown in Fig. 8.8 is applied to the circuit of Fig. 8.1. The effect upon the circuit can be considered to be a combination of the conditions in the previous two sections.

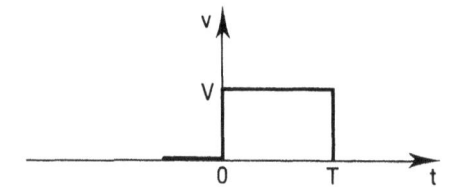

Figure 8.8

For the period from 0 to T seconds, the current in the circuit and the voltages across the resistor and the inductor will be given by Equations (8.2), (8.5) and (8.6), respectively, while for the period following T, these quantities are given by Equations (8.9), (8.10) and (8.11), respectively. The corresponding waveform of current may be obtained by combining Fig. 8.3 with Fig. 8.6(a), putting $0 \geq 5\tau$ in the latter. Those of v_R and v_L can be obtained by combining Fig. 8.4 with Fig. 8.6(b), again changing the origin of the latter to 5τ. When you have done this you should obtain graphs similar to those of Figs 8.10 and 8.12.

The RL integrator circuit

If the output voltage of the RL circuit shown in Fig. 8.9 is taken to be the voltage across the resistor, then the circuit is called an integrator because the output waveform approximates to the integral of the input voltage.

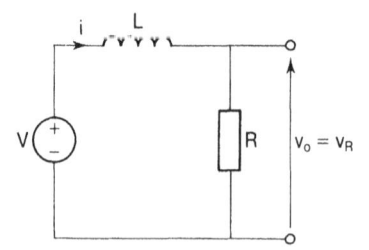

Figure 8.9

Example 8.4

A pulse of magnitude 5 V and width 40 μs is applied to an integrator circuit consisting of an inductor of 8 mH in series with a resistor of 2 kΩ. Draw the waveforms of the output voltage, v_o, and the current, i.

Solution

The circuit is as shown in Fig. 8.9 with $R = 2$ kΩ, $L = 8$ mH and $V = 5$ V. First consider the period 0–T (0–40 μs). The time constant of the circuit is

$$\tau = L/R = 8 \times 10^{-3}/2 \times 10^3 = 4 \times 10^{-6}\,\text{s} = 4\,\mu\text{s}$$

Now $5\tau = 20$ μs and since the pulse width is twice as long as this the steady state conditions will have been reached well before the pulse is removed.

The steady state value of the current is $V/R = 5/2 \times 10^3 = 2.5 \times 10^{-3}$ A = 2.5 mA and this is considered to have been reached in 20 μs. The instantaneous value of the current is given by Equation (8.2) to be $i = I[1 - \exp(-Rt/L)]$. After the time constant (4 μs) the current will have reached $0.632\,I = 0.632 \times 2.5 \times 10^{-3}$ A = 1.58 mA. The output voltage is given by Equation (8.5) to be $v_o = v_R = V[1 - \exp(-Rt/L)]$ and after 4 μs this will also have reached 63.2 per cent of the steady state value (5 V). The output voltage will then be $0.632 \times 5 = 3.16$ V.

Next we consider the period subsequent to the removal of the pulse (i.e. after 40 μs). During this period the current will be decaying in accordance with Equation (8.9) while the output voltage is also decaying according to Equation (8.10). After $5\tau = 20$ μs (i.e. a total of 60 μs after $t = 0$), both will have died away to zero. After $\tau = 4$ μs (44 μs after $t = 0$) the current will have reached $I\exp(-1)$ so that $i = 0.368\,I = 0.368 \times 2.5 = 0.92$ mA. Similarly $v_o = v_R = 0.368\,V = 0.368 \times 5 = 1.84$ V.

The graphs of current and output voltage are now drawn as shown in Fig. 8.10.

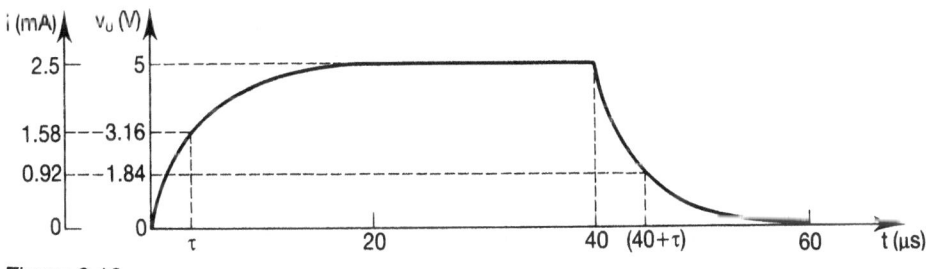

Figure 8.10

The RL differentiator circuit

If the output voltage is taken to be that across the inductor as shown in Fig. 8.11 the circuit is called a differentiator because the output voltage waveform approximates to the differential of the input voltage waveform.

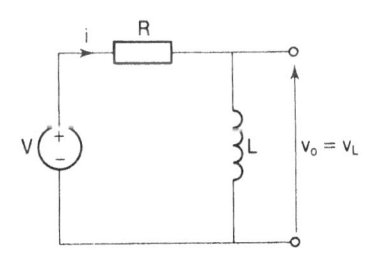

Figure 8.11

Example 8.5

Derive the waveforms of the current and the output voltage for the differentiator circuit given in Fig. 8.11 when $R = 2\ \text{k}\Omega$, $L = 8\ \text{mH}$ and the input voltage pulse is 5 V in magnitude with a width of 40 μs.

Solution

This is the same circuit as that of the previous example simply reconfigured to make v_L the output voltage. The time constant of the circuit is therefore the same as before at 4 μs,

The pulse width is ten times the time constant of the circuit so that, following the application of the pulse, the current and both element voltages will reach their final steady state values well before it is subsequently removed.

Consider the period 0–40 μs. The current waveform is identical to that derived for the previous example. The output voltage v_o is now the voltage across the inductor v_L and this is given by Equation (8.6) as

$$v_L = V \exp(-Rt/L)$$

After $t = \tau (4\ \mu s)$

$$v_L = 5 \exp(-1) = 0.368 \times 5 = 1.84\ \text{V}$$

After $t = 5\tau (20\ \mu s)$

$$v_L = V = 5\ \text{V}$$

Consider now the period following the removal of the pulse, i.e. after 40 μs. Again the current waveform is the same as that obtained in Example 8.4. The output voltage is given by Equation (8.11) to be $v_L = -V \exp(-Rt'/L)$ where $t' = (t - 40)\ \mu s$.

At $t = 40\ \mu s$, $t' = 0$ and $v_L = -V \exp(0) = -V$
At $t = 44\ \mu s$, $t' = 4\ \mu s$ and $v_L = -V \exp(-1) = -0.368 \times 5 = 1.84\ \text{V}$
At $t = 60\ \mu s$, $t' = 20\ \mu s$ and $v_L = -V \exp(-5) = 0$

The waveforms of current and output voltage may now be drawn as in Fig. 8.12.

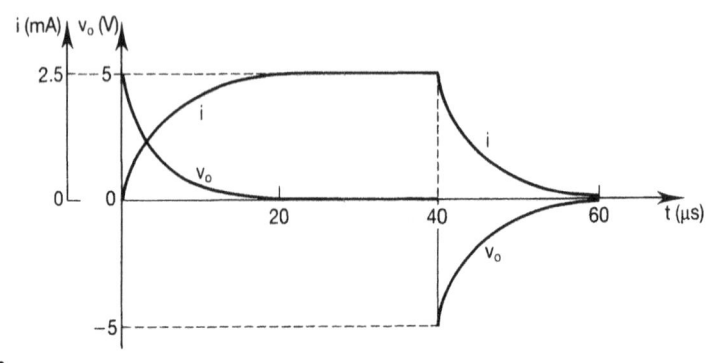

Figure 8.12

8.3 CIRCUITS CONTAINING RESISTANCE AND CAPACITANCE

The sudden application of a step voltage

Let the step function shown in Fig. 8.2 be applied to the circuit of Fig. 8.13. Applying KVL we see that

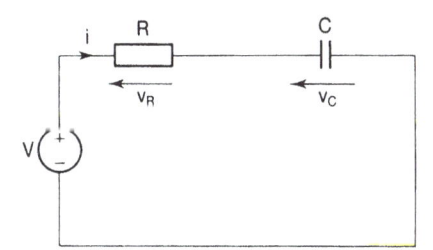

$$V = v_R + v_C \tag{8.12}$$

Figure 8.13

Now $v_R = iR$ and from Equation (2.1), Chapter 2, $i = dq/dt$ where i is the current through the capacitor and q is its charge at any instant. Also, from Equation (2.18), $q = Cv_C$ where C is the capacitance of the capacitor and v_C is the potential difference between its plates at any instant, so that $i = Cdv_C/dt$. Therefore $v_R = iR = RCdv_C/dt$ and Equation (8.12) becomes

$$V = RCdv_C/dt + v_C \tag{8.13}$$

Rearranging, we have

$$V - v_C = RCdv_C/dt$$

Separating the variables,

$$dv_C/(V - v_C) = dt/(RC)$$

Integrating we get

$$\ln(V - v_C) = t/(RC) + C$$

where C is the constant of integration. Taking antilogs we have

$$V - v_C = \exp[(-t/RC) + C]$$
$$= \exp(-t/RC)\exp(C)$$

When $t = 0$, $v_C = 0$ and so $\exp(C) = V$. Therefore

$$V - v_C = V\exp(-t/RC) \quad \text{and} \quad v_C = V - V\exp(-t/RC)$$

Finally

$$v_C = V[1 - \exp(-t/RC)] \tag{8.14}$$

This describes an exponential growth of voltage and is shown in Fig. 8.14.

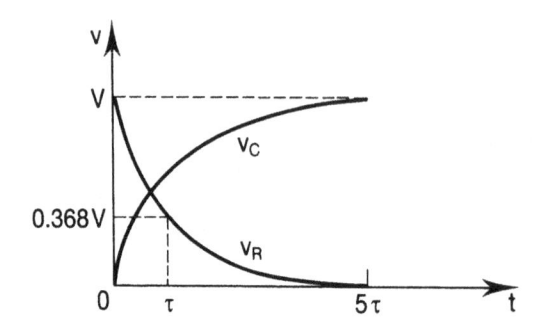

Figure 8.14

The time constant of an RC circuit

The product RC is called the time constant (symbol τ) of the RC circuit and it is measured in seconds. When $t = RC$,

$$v_C = V[1 - \exp(-1)] = 0.632\,V$$

After a time equal to the time constant, therefore, the voltage across the capacitor will have reached 63.2 per cent of its final steady state value V. Compare this with the growth of current in an RL circuit when subjected to a step input (Equation (8.3)).

After a period equal to $5RC$, $v_C = V[1 - \exp(-5)] = 0.993$ V. To all intents and purposes, therefore, the voltage across the capacitor will have reached its steady state value after 5τ seconds. The capacitor is then said to be fully charged and there is no longer any movement of charge to its plates. The current in the circuit is therefore zero and the voltage across the resistor is also zero. Because the voltage across the capacitor is initially zero, the whole of the applied voltage V appears across the resistor in accordance with KVL so $V = iR$ and the current immediately jumps to the value $I = V/R$. Thereafter it may be found from

$$i = C\,dv_C/dt = C(d/dt)\{V[1 - \exp(-t/CR)]\}$$
$$= CV[0 - (-1/RC)\exp(-t/RC)] = (V/R)\exp(-t/RC)$$

Finally, since $V/R = I$, then

$$i = I\exp(-t/RC) \tag{8.15}$$

This shows that, as the charging process proceeds, the current decays exponentially from its initial value of $I(=V/R)$ to zero. The waveform of the current is shown on the following page in Fig. 8.15.

The voltage across the resistor, v_R is given by $iR = IR\exp(-t/CR)$ and, since $IR = V$, then

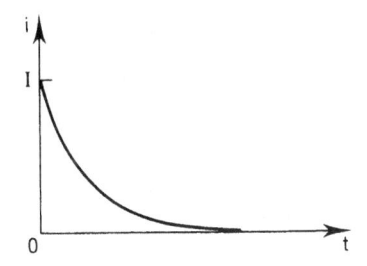

Figure 8.15

$$v_R = V \exp(-t/CR) \tag{8.16}$$

This describes an exponential decay of voltage and is shown in Fig. 8.14.

Example 8.6

In the circuit of Fig. 8.16, the switch S is closed at $t = 0$. Determine the value of the current (i_s) drawn from the supply after 30 ms.

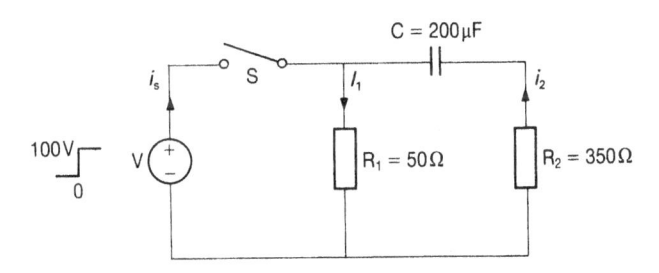

Figure 8.16

Solution

From Equation (8.15) the current through the branch containing C and R_2 is given by

$$i_2 = I_2 \exp(-t/CR_2)$$

The time constant of this branch is

$$\tau = CR_2 = 200 \times 10^{-6} \times 350 = 0.07 \text{ s}$$

The steady state value of the current through the branch is

$$I_2 = V/R_2 = 100/350 = 0.286 \text{ A}.$$

After 30 ms,

$$i_2 = 0.286 \exp(-0.03/0.07) = 0.186 \text{ A}$$

The current $I_1 = V/R_1 = 100/50 = 2$ A and it reaches this value immediately S is closed because the branch is purely resistive. After 30 ms, therefore,

$$i_s = I_1 + i_2 = 2 + 0.186 = 2.186 \text{ A}$$

The discharge of a capacitor

Once the capacitor in Fig. 8.13 has been charged to V volts, it will remain at that voltage so long as the input voltage remains at V volts. Even if the supply is disconnected, as shown in Fig. 8.17(a), the capacitor (provided that it has no leakage resistance between its plates) will remain charged to V volts. If, however, the terminals A and B are now short circuited as shown in Fig. 8.17(b) the capacitor will begin to release its stored energy. The current will now be *leaving* its positive plate.

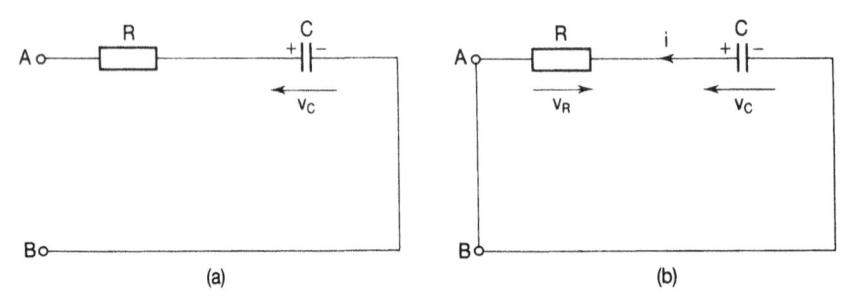

Figure 8.17

Applying KVL to the circuit of Fig. 8.17(b) we see that

$$v_C - iR = 0 \tag{8.17}$$

Now $i = -(dq/dt)$, the minus sign indicating that the charge on the plates is decaying. Since $Q = CV$ then $q = Cv_C$ and

$$-(dq/dt) = -C(dv_C/dt)$$

Substituting in Equation (8.17) we get

$$v_C - [-C(dv_C/dt)]R = 0$$

$$v_C = -CR(dv_C/dt)$$

Separating the variables we have $dv_C/v_C = -dt/CR$. Integrating gives

$$\ln v_C + C = -t/CR \tag{8.18}$$

where C is the constant of integration. At $t = 0$, $v_C = V$ so that $\ln V + C = 0$ and $C = -\ln V$. Substituting in Equation (8.18) we obtain

$$\ln v_C - \ln V = -t/CR \Rightarrow \ln(v_C/V) = -t/CR$$

Taking antilogs

$$v_C/V = \exp(-(t/CR))$$

Finally

$$v_C = V \exp(-t/CR) \tag{8.19}$$

The current at any instant is given by

8.3 Circuits containing resistance and capacitance

$i = v_C/R = (V/R) \exp(-t/CR)$

and since $V/R = I$ then

$i = I \exp(-t/CR)$ (8.20)

The voltage across the resistor decays as the current through it decays so that $v_R = iR$ in magnitude. Thus

$v_R = RI \exp(-t/CR) = V \exp(-t/CR)$

Also, $v_R + v_C = 0$ so that the voltage across the resistor is given by

$v_R = -v_C = -V \exp(-t/CR)$ (8.21)

The waveforms described by Equations (8.19), (8.20) and (8.21) are shown in Fig. 8.18(a), (b) and (a), respectively. Remember, i is discharging current.

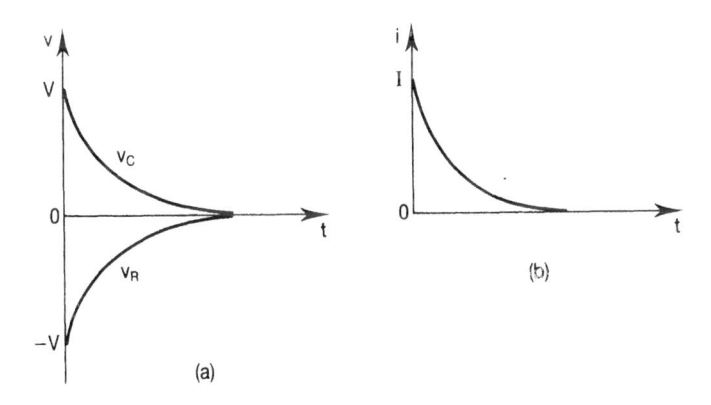

Figure 8.18

Example 8.7

A 10 μF capacitor is charged to 150 V and then allowed to discharge through its own leakage resistance. After 200 s, it is observed that the voltage, measured on an electrostatic voltmeter, has fallen to 90 V. Calculate the leakage resistance of the capacitor.

Solution

From Equation (8.19) we have that $v_C = V \exp(-t/CR)$. In this case R is the leakage resistance of the capacitor. The resistance of the electrostatic voltmeter which is in parallel with R can be taken to be infinite. Putting in the values we have $90 = 150 \exp[-200/(10 \times 10^{-6} \times R)]$, so

$-200/(10 \times 10^{-6} \times R) = \ln(90/150)$

and

$R = -200/[10 \times 10^{-6} \times \ln(90/150)] = 39.15 \, \text{M}\Omega$

Response of the RC circuit to a pulse input voltage

Let the pulse of amplitude V and width T ($>5\tau$) shown in Fig. 8.8 be applied to the circuit of Fig. 8.13. From $t = 0$ to $t = \tau$ the conditions are similar to those when a step input is applied and Equations (8.14), (8.15) and (8.16) apply. From $t = 5\tau$ to $t = T$, the capacitor is fully charged so that $v_C = V$, $i = 0$ and $v_R = 0$. After the removal of the pulse at $t = T$ the capacitor discharges and Equations (8.19), (8.20) and (8.21) apply. These points are illustrated in the waveforms of Fig. 8.19.

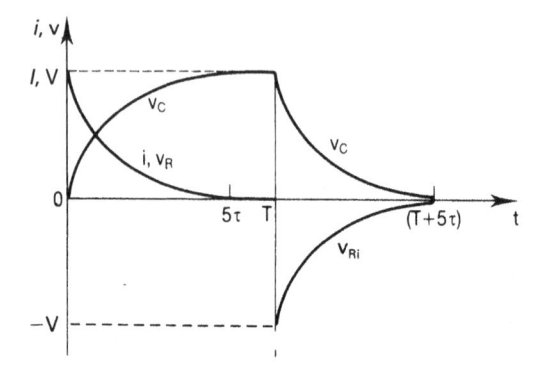

Figure 8.19

The RC integrator circuit

If the output voltage is taken across the capacitor as shown in the circuit of Fig. 8.20, we have an integrator circuit, so called because the output voltage approximates to the integral of the input voltage.

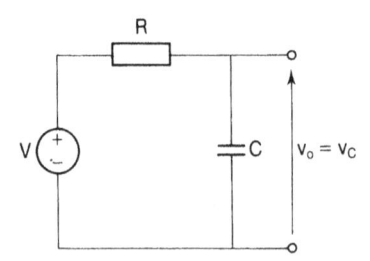

Figure 8.20

Single pulse input

Following the application of the pulse's leading edge to the input terminals, the capacitor charges up in accordance with Equation (8.14) so that the capacitor voltage is given by

$$v_C = V[1 - \exp(-t/CR)]$$

If the pulse width T is greater than 5τ the capacitor will be fully charged before the pulse is removed. After the pulse is removed, the capacitor discharges in accordance with Equation (8.19). The output voltage waveform is then as shown in Fig. 8.21.

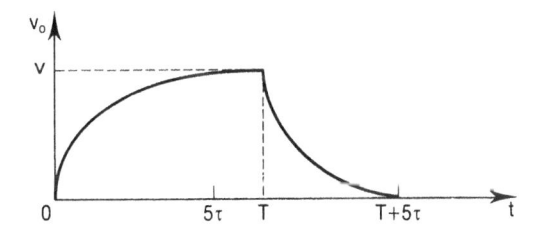

Figure 8.21

Multiple pulse input

If a train of pulses, for which the pulse width T ($>5\tau$) is equal to the pulse separation, is applied to the input terminals then the waveform shown in Fig. 8.21 will repeat. The capacitor will be fully charged before the end of each pulse and subsequently fully discharged before the arrival of the next pulse. This is shown in Fig. 8.22.

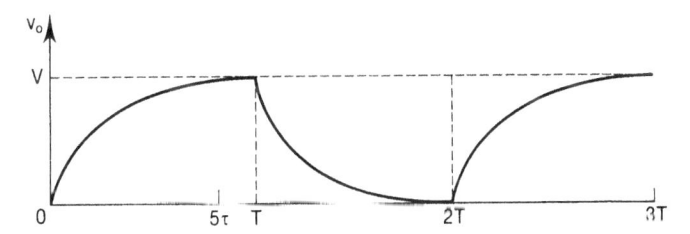

Figure 8.22

Pulses for which $T < 5\tau$

If the width of the pulse is less than five time constants then the capacitor will not be fully charged before the pulse is removed (i.e. v_C will be only a fraction of V). The smaller the pulse width, the smaller the fraction of V reached when the pulse is removed. However, the time constant CR remains unchanged so that it takes the same time for the capacitor to become fully discharged after the pulse is removed, regardless of the pulse width. This is shown in Fig. 8.23, which is drawn for a pulse width $T = 2\tau$.

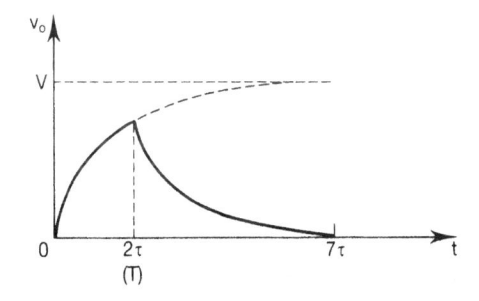

Figure 8.23

Example 8.8

The integrator circuit shown in Fig. 8.20 has $R = 10 \text{ k}\Omega$ and $C = 0.01 \text{ μF}$. Sketch the waveform of the output voltage v_o for each of the following input conditions:

1 the input signal is a single pulse of amplitude 10 V and width $T = 600 \text{ μs}$;

2 the input is a train of pulses of amplitude 10 V, width 100 μs and separation 500 μs.

Solution

The time constant of the circuit is given by
$$\tau = CR = 0.01 \times 10^{-6} \times 10 \times 10^3 = 100 \times 10^{-6} \text{ s}$$

1 The pulse width is greater than five time constants so that the capacitor voltage v_c will reach its steady state value of 10 V before the pulse is removed. It will remain at this value until the pulse is removed at $t = 600 \text{ μs}$, after which it will decay to zero as shown in Fig. 8.24.

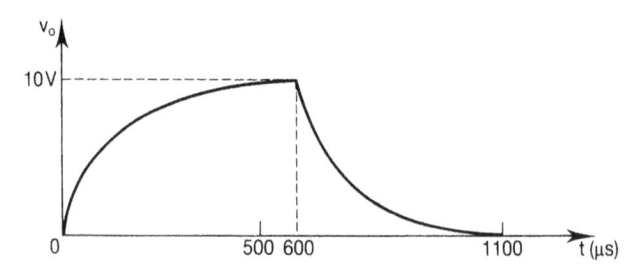

Figure 8.24

2 The pulse width is 100 μs which is less than five time constants (in fact it is one time constant) so that the capacitor will only be partially charged before a pulse is removed. After 100 μs,

$$v_o = v_C = V(1 - \exp(-t/CR)) = 10\{1 - \exp[-100/(0.01 \times 10 \times 10^3)]\}$$
$$= 6.32 \text{ V}$$

The pulse separation is 500 μs which is 5τ so the capacitor will be fully discharged just as the next pulse in the train arrives at the input terminals. The waveform of v_o is given in Fig. 8.25.

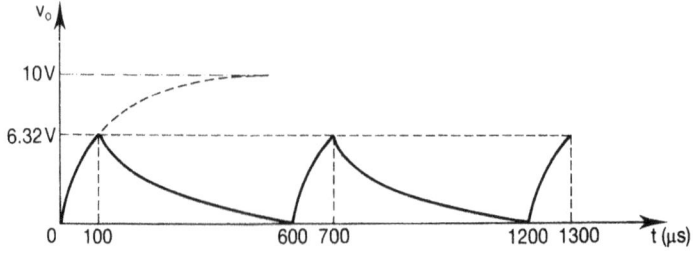

Figure 8.25

The RC differentiator circuit

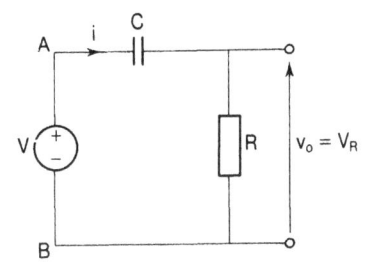

Figure 8.26

If the output voltage is taken across the resistor as shown in Fig. 8.26, then the circuit is called a differentiator. At instant $t = 0$, when the leading edge of the pulse arrives at the input terminals AB, the voltage across the capacitor cannot suddenly change so the potential difference across it is zero and a voltage of $+V$ appears at the output terminals. Thus v_o ($=v_R$) $= V$. The capacitor then begins to charge in accordance with Equation (8.14) so $v_C = V[1 - \exp(-t/CR)]$. At the same time the potential difference across the resistor falls exponentially in accordance with Equation (8.16).

If the pulse width is greater than five time constants ($5CR$) then the voltage V_o will have reached zero before the trailing edge of the pulse arrives (i.e. before the pulse is removed). At this point $v_{in} = 0$ so that $v_C + v_R = 0$ and because $v_C = V$ and cannot change suddenly then v_R must immediately become equal to $-V$.

As the capacitor then begins to discharge in accordance with Equation (8.19), v_R will rise towards zero in accordance with Equation (8.21) so that at every instant $v_C + v_R = 0$. The waveforms of v_C and v_R ($=v_o$) are therefore as shown in Fig. 8.27.

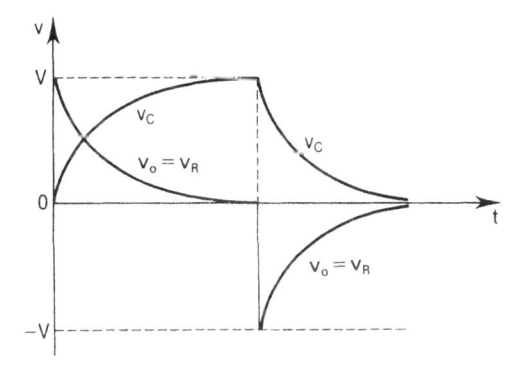

Figure 8.27

If the pulse width is less than five time constants then the capacitor will only be partially charged by the time the pulse is removed, so that v_C will be less than V and v_R will not have fallen to zero. The input voltage V_{in} is now equal to zero so that $V_C + V_R = 0$. The voltage across the resistor will therefore have to change immediately to $-v_C$ at that instant. The voltage across the capacitor will now begin to decay to zero in accordance with Equation (8.19) and will take a time equivalent to five time constants to do so. In the same time the voltage across the resistor will rise to zero. The waveforms are thus as shown in Fig. 8.28, which is drawn for a pulse width $T = 2\tau$.

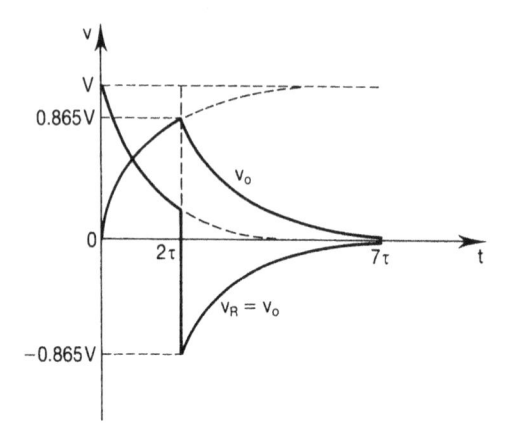

Figure 8.28

Example 8.9

The differentiator circuit shown in Fig. 8.26 has $R = 2$ MΩ and $C = 1.25$ pF. Obtain the waveform of the output voltage v_o for each of the following input conditions:

1. a single pulse of amplitude 10 V and width 15 μs;
2. a train of pulses, each of amplitude 10 V and width 5 μs, separated by 15 μs.

Solution

1. When the pulse is applied to the circuit, $v_C + v_R = 10$ V and all of this appears across the resistor because the capacitor voltage cannot change instantaneously. The output voltage v_o therefore immediately becomes equal to V_{in} (=10 V). The time constant of the circuit is

$$\tau = CR = 1.25 \times 10^{-12} \times 2 \times 10^6 = 2.5 \text{ μs}$$

The pulse width is 15 μs and since this is $>5\tau$ then the capacitor voltage will have reached V (=10 V) and the output voltage v_o (=v_R) will have reached zero before the pulse is removed. The output voltage will remain at zero until the trailing edge of the pulse arrives at the input terminals. During this time the capacitor voltage $v_C = 10$ V. When the pulse is removed the input voltage is zero and so $v_C + v_R = 0$. The capacitor voltage cannot suddenly change but the voltage across the resistor can and it changes immediately to −10 V.

The output voltage waveform can now be drawn as shown in Fig. 8.29.

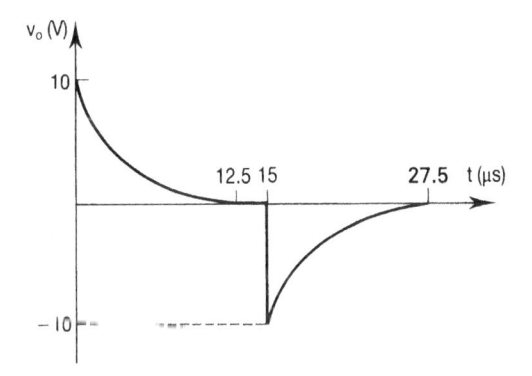

Figure 8.29

2 Again the output voltage will immediately jump to 10 V as the first pulse arrives at the input terminals, for the reason given in part (1). In this case the pulse width is 5 μs which is <5τ which means that the capacitor will be only partially charged by the time the trailing edge of any particular pulse arrives at the input terminals. Also the output voltage will have fallen to a value >0 and in fact will be given by

$v_o = V \exp(-t/CR) = 10 \exp(-5/2.5) = 1.35$ V

At this instant $v_i = 0$ so that $(v_C + v_R) = 0 \Rightarrow v_R = -v_C$. But

$v_C = (10 - v_o) = (10 - 1.35)$ V $= 8.65$ V

so

$v_R = v_o = -8.65$ V

The voltage across the resistor subsequently rises towards zero. Since the next pulse in the train arrives after 15 μs, which is longer than five time constants, the output voltage becomes zero before it arrives at the input terminals. The waveform of v_o may now be drawn as shown in Fig. 8.30.

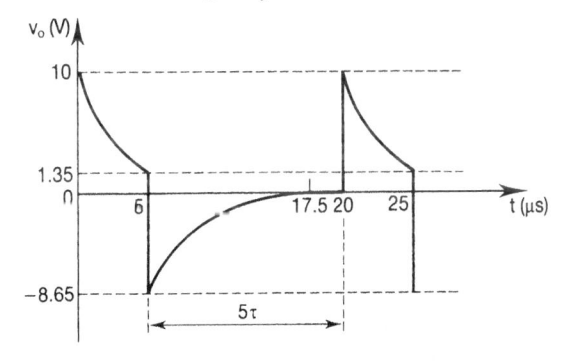

Figure 8.30

8.4 THE LAPLACE TRANSFORM

It is a common technique to transform problems to a different form in order to make their solution easier, even if the resulting processes are longer. One example is the use of logarithms to transform the process of multiplication or division into the simpler one of addition or subtraction. The method is to:

1 look up the logarithms of the numbers to be multiplied (or divided);

2 add (or subtract) the logarithms;

3 look up the antilogarithm of the result in order to obtain the answer to the original problem.

The equations associated with the transient operation of electric circuits are differential equations in the time domain. A stimulus which is a function of time is applied to the circuit whose behaviour is then described by one or more differential equations. These equations then have to be solved in order to determine the response of the circuit to the stimulus. By means of the Laplace transform it is possible to convert these differential equations into algebraic equations involving a complex variable, s. After manipulation in order to solve these algebraic equations (which are easier to solve than differential equations) the inverse transform is found, which gives the time response to the original stimulus. The method, then, is to:

1 set up the differential equations which describe the operation of the circuit;

2 look up the table of Laplace transforms in order to convert these to algebraic equations;

3 solve the algebraic equations to find the response to the circuit in terms of the complex variable, s;

4 look up the table of inverse transforms to find the time response of the circuit to the original stimulus.

The Laplace transform is named after Pierre-Simon Laplace, a French mathematician. It is written as

$$L[f(t)] = F(s) \tag{8.22}$$

which is read 'the Laplace transform of the function of time $f(t)$ is equal to a function of s'. It is defined as

$$L[f(t)] = \int_0^\infty f(t) \exp(-st)\, dt \tag{8.23}$$

Thus the original function $f(t)$ is first multiplied by the exponential decay $\exp(-st)$ and the result is integrated from zero to infinity. The value of s must be such that Equation (8.23) converges to zero as $t \to \infty$ (it almost always does in problems associated with electrical circuits).

The inverse transform is written

$$f(t) = L^{-1}[F(s)] \qquad (8.24)$$

$F(s)$ is often written as $\bar{f}$ which is read 'f bar'.

The Laplace transforms of many different functions of time have been determined and tabulated. Since these tables are readily available there is no need to derive the required transform in any particular problem. However, a few are derived here in order to show the method.

The Laplace transform of the exponential function

If $f(t) = \exp(-at)$ then

$$L[f(t)] = F(s) = \int_0^\infty \exp(-at)\exp(-st)\,dt = \int_0^\infty \exp[-(s+a)t]\,dt$$
$$= [-1/(s+a)\{\exp(-(s+a))t\}]_0^\infty$$
$$= 1/(s+a) \qquad (8.25)$$

The Laplace transform of the unit step function

If $f(t) = 1$ for $t \geq 0$ and $f(t) = 0$ for $t < 0$ then

$$L[f(t)] = F(s) = \int_0^\infty 1\exp(-st)\,dt = -(1/s)[\exp(-st)]_0^\infty$$
$$= 1/s. \qquad (8.26)$$

It follows that, for a step function of amplitude A, the Laplace transform is A/s.

The Laplace transform of the derivative of a function

The derivative of the function $f(t)$ is $d(f(t))/dt$. Let this be denoted by $f'(t)$. Then

$$L[f'(t)] = \int_0^\infty f'(t)\exp(-st)\,dt$$

Integrating by parts ($\int u\,dv = [uv] - \int v\,du$) we have, with $u = \exp(-st)$ and $dv = f'(t)\,dt$ (so that $v = f(t)$),

$$\int_0^\infty f'(t)\exp(-st)\,dt = [\exp(-st)f(t)]_0^\infty - \int_0^\infty f(t)\{-s\exp(-st)\}\,dt$$

The upper limit of the first term on the right-hand side must be zero in order that it tends to infinity as t tends to zero. Thus the right-hand side becomes

$$-f(0) + s\int_0^\infty f(t)\exp(-st)\,dt$$

and the second term is s times the Laplace transform of the original function of t

$$\therefore L[f'(t)] = sL[f(t)] - f(0) \qquad (8.27)$$

These transform pairs, together with a few other commonly encountered ones, are given in Table 8.1.

Table 8.1

	$f(t)$	Description	$L[f(t)] = F(s)$
1	$\exp(-\alpha t)$	exponential function	$1/(s + \alpha)$
2	1	unit step function	$1/s$
3	A	step function of amplitude A	A/s
4	$d\{f(t)\}/dt$	differential of $f(t)$	$sF(s) - f(0)$
5	$\int_0^t f(t)dt$	integral of $f(t)$	$(1/s)F(s) + f(0)/s$
6	t	ramp function	$1/s^2$
7	$\sin \omega t$	sinusoidal function	$\omega/(s^2 + \omega^2)$
8	$\cos \omega t$	cosinusoidal function	$s/(s^2 + \omega^2)$
9	$\exp(-\alpha) \sin \omega t$	exponentially decaying sinusoidal function	$\omega/[(s + \alpha)^2 + \omega^2]$
10	$\exp(-\alpha t) \cos \omega t$	exponentially decaying cosinusoidal function	$(s + \alpha)/[(s + \alpha)^2 + \omega^2]$

Application to electrical circuit transient analysis

For transient circuit analysis we convert the original circuit into a transform circuit, and to do this we need to be able to transform the circuit elements R, L and C.

Resistance

If a resistor R has a current flowing through it given by $i = f(t)$ then the voltage across it will also be a function of time, given by $v(t) = Ri(t)$. Taking Laplace transforms we have

$L[v(t)] = RL[i(t)]$
$v(s) = RI(s)$

Thus resistance is the same in the s-domain as it is in the time domain.

Inductance

If an inductor L has a current flowing through it given by $i(t)$ then the voltage across it will be given by $v(t) = Ldi/dt$. Taking Laplace transforms we have, from number 4 in Table 8.1.

$v(s) = Lsi(s) - Li(0)$

If the current is initially zero the second term on the right-hand side disappears ($Li(0) = 0$)

Capacitance

If the current is given by $i(t)$ then the voltage across the plates will be given by $v_C(t) = (1/C)\int i\, dt$. Taking Laplace transforms and using number 5 from the table we have

$$v_C(s) = (1/Cs)i(s) + v_C(0)/s$$

If the capacitor is initially uncharged then $v_C(0) = 0$ and the second term on the right-hand side disappears.

Example 8.10

Obtain the transform circuit for the circuit shown in Fig. 8.31

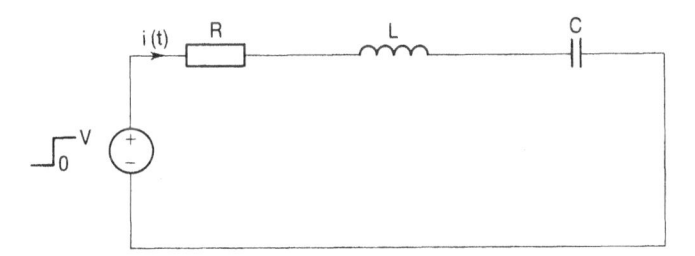

Figure 8.31

Solution

The voltage source is a step function of amplitude V. From Table 8.1 we see that this transforms to V/s (pair number 3 with $A = V$).

The resistor R in the original circuit remains unchanged in the transform circuit.

The inductor L transforms to an element Ls in series with a source $Li(0)$. This source is short circuited if the current is initially zero.

The capacitor C transforms to an element $1/Cs$ in series with a source $v_C(0)/s$. This source is short circuited if the capacitor is initially uncharged.

The transform circuit therefore takes the form shown in Fig. 8.32.

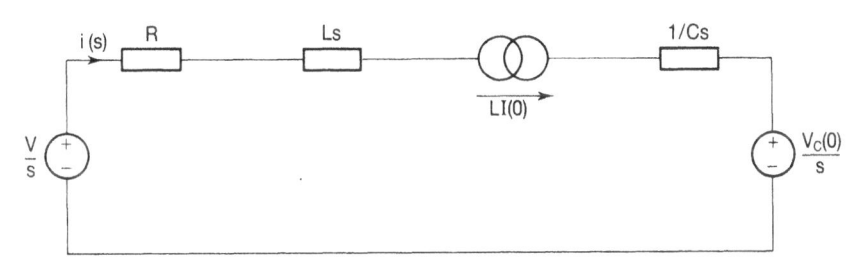

Figure 8.32

Transforming to the time domain from the s-domain

From the transform circuit of Fig. 8.32 we have that

$V/s + Li(0) - V_C(0)/s = i(s)[R + Ls + 1/Cs]$

Assuming that the initial conditions are zero ($i = 0$ and $v_C = 0$) then $Li(0) = 0$ and $V_C(0)/s = 0$. Thus

$i(s) = V/s/[R + Ls + 1/Cs]$

In order to proceed from here to obtain i as a function of time we have to use the inverse Laplace transform table. Very often the expression for $i(s)$ is of the form $i(s) = M(s)/N(s)$ where $M(s)$ and $N(s)$ are polynomials. These must be reduced to a series of partial fractions in order to identify a suitable transform pair from the table.

Partial fractions

By using a common denominator the expression $[A/(s + 1)] + [B/(s + 2)]$ can be written $\{A(s + 2) + B(s + 1)\}/\{(s + 1)(s + 2)\}$. It follows that the inverse process is possible so that the expression with a common denominator may be converted to a series of terms with separate denominators. These are called partial fractions. There are a number of techniques for finding the partial fractions of an expression, one of which is known as 'equating coefficients'. The following example illustrates the method.

Example 8.11

Find the time response of a circuit for which the current in the transform circuit is given by $i(s) = (V/R)(s + 3)/[(s + 1)(s + 2)]$.

Solution

To identify suitable transform pairs we must find the partial fractions of $(s + 3)/[(s + 1)(s + 2)]$

Let

$(s + 3)/[(s + 1)(s + 2)] = [A/(s + 1)] + [B/(s + 2)]$

Since

$[A/(s + 1)] + [B/(s + 2)] = [A(s + 2) + B(s + 1)]/[(s + 1)(s + 2)]$

then

$s + 3 \equiv A(s + 2) + B(s + 1)$

i.e.

$1s^1 + 3s^0 \equiv (A + B)s^1 + (2A + B)s^0$ remembering that $s^0 = 1$.

Equating the coefficients of s^1 we have that

$1 = A + B$

Equating the coefficients of s^0 we have that

$3 = 2A + B$

By subtraction, $A = 2$; by substitution, $1 = 2 + B \Rightarrow B = -1$. Therefore

$i(s) = [2/(s + 1)] - [1/(s + 2)]$

From the Table 8.1 we see that the inverse transform of $2/(s + 1)$ is $2 \exp(-t)$. This is from pair number 1 with $\alpha = 1$. Using the same inverse pair with $\alpha = 2$ we see that the inverse transform of $1/(s + 2)$ is $\exp(-2t)$. Thus the time response of the circuit is given by $i(t) = (V/R)[2 \exp(-t) - \exp(-2t)]$.

Example 8.12

Obtain an expression for the current in the circuit of Fig. 8.33 following the closing of switch S.

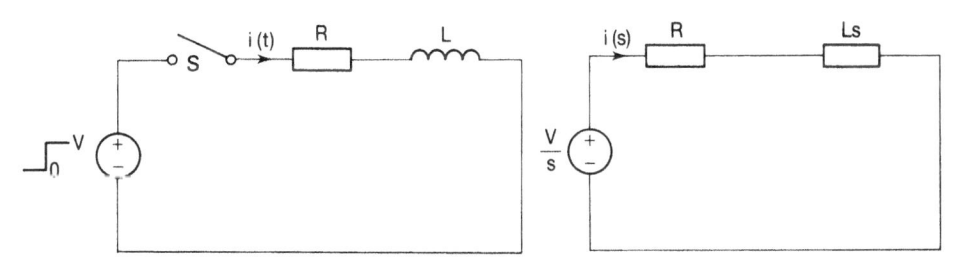

Figure 8.33

Figure 8.34

Solution

The circuit equation is $V = iR + L\,di/dt$, and earlier in the chapter we solved this equation for i and obtained the result shown in Equation (8.2). We will now obtain the same result using the Laplace transform method. First we obtain the transform circuit.

At the instant of closing the switch S, the current is zero so that there is no $LI(0)$ source and the transform circuit takes the form shown in Fig. 8.34.

The closing of the switch is equivalent to there being a step function of amplitude V applied, the transform of which is V/s (pair 3 from Table 8.1 with $A = V$). From this circuit we see that

$i(s) = (V/s)/(R + Ls) = V/s[R(1 + (L/R)s)]$

Multiplying numerator and denominator by R/L we have

$i(s) = V(R/L)/s[R(1 + (L/R)s)][R/L] = V(R/L)/Rs[s + (R/L)]$
$= (V/R)\{(R/L)/[s(s + (R/L)]\}$

198 *Transient analysis*

Let

$$(R/L)/s[s + (R/L)] = [A/s] + [B/\{s + (R/L)\}]$$

Then

$$R/L \equiv A[s + (R/L)] + Bs$$

so that

$$(R/L)s^0 \equiv (A + B)s + (AR/L)s^0$$

Equating the coefficients of s we have that $0 = A + B \Rightarrow B = -A$. Equating the coefficients of s^0 we have that $R/L = AR/L \Rightarrow A = 1$ and $B = -A = -1$. Thus

$$(V/R)\{(R/L)/[s\{s + (R/L)\}]\} = (V/R)\{[1/s] - [1/(s + (R/L))]\}$$

From Table 8.1 we see that $1/s$ is the transform of 1 (transform pair number 1) and that $1/[s + (R/L)]$ is the transform of $\exp(-Rt/L)$ (transform pair number 1 with $\alpha = R/L$). In the time domain, therefore, the current is given by $i(t) = (V/R)\{1 - \exp(-Rt/L)\}$ and since $V/R = I$,

$$i(t) - I[1 = \exp(-Rt/L)]$$

This is the same result as that obtained in Equation (8.2).

Example 8.13

In the circuit shown in Fig. 8.35, the capacitor is initially charged to 20 V when the switch is closed at an instant $t = 0$. Obtain an expression for the voltage (v_C) across the capacitor as a function of time. Hence determine the time taken for v_C to double its initial value.

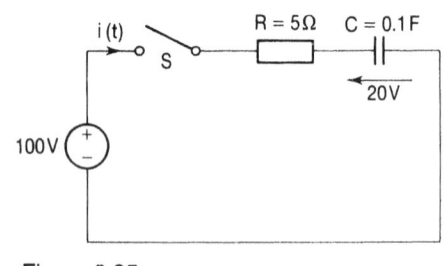

Figure 8.35

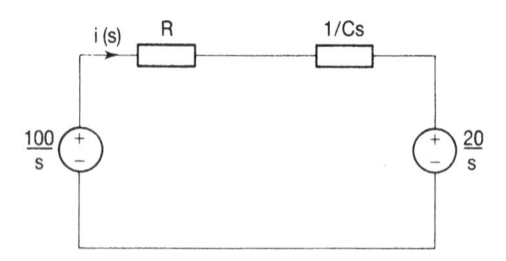

Figure 8.36

Solution

First we obtain the transform circuit. The step voltage of amplitude 100 V resulting from the closing of the switch S transforms to $100/s$ (Table 8.1, transform pair number 3 with $A = 100$). The capacitor is initially charged to 20 V so that there will be a source $V_C(0)/s$ ($=20/s$) in series with the element

$1/Cs$ (=$1/0.1s$). The transform circuit thus takes the form shown in Fig. 8.36. From this circuit we see that

$$i(s) = [(100/s) - (20/s)]/[R + (1/Cs)] = 80/s[R + (1/Cs)]$$

Now

$$v_C(s) = i(s)(1/Cs) = 80/\{s[R + (1/Cs)]Cs\} = 80/s(CRs + 1)$$

Multiplying numerator and denominator by $1/CR$ we get

$$v_C(s) = 80(1/CR)/s[s + (1/CR)]$$

Let

$$(1/CR)/s[(s + 1/CR)] = [A/s] + [B/\{s + (1/CR)\}]$$

Then

$$1/CR = A[s + (1/CR)] + Bs \equiv (A + B)s + A/CR$$

Equating the coefficients of s we have that $A + B = 0 \Rightarrow A = -B$. Equating the coefficients of s^0 we have that $1/CR = A/CR \Rightarrow A = 1$ and $B = -A = -1$.

Thus

$$v_C(s) = 80[1/s - 1/(s + 1/CR)]$$

From Table 8.1 we see that $1/s$ is the transform of 1 and that $1/(s + 1/CR)$ is the transform of $\exp(-t/CR)$ (transform pair number 1 with $\alpha = 1/CR$). In the time domain, therefore

$$v_C(t) = 80[1 - \exp(-t/CR)] = 80[1 - \exp(-t/0.5)]$$

When v_C has reached 40 V (i.e. it is double its initial value) we have

$$40 = 80[1 - \exp(-t/0.5)]$$

$$\exp(-2t) = 0.5$$

Therefore

$$-2t = \ln 0.5 \quad \text{and} \quad t = 0.345 \text{ s}$$

Example 8.14

An inductor of 250 µH inductance is energized from a 1 kV d.c. supply via two thyristors connected in series. The thyristors are identical except that one of them has a delay at turn-on of a few microseconds longer than the other. Voltage sharing is assisted by placing an RC circuit in parallel with each thyristor. Obtain an expression for the current in the RC circuit connected across the 'slow' thyristor after the other one has turned on.

Solution

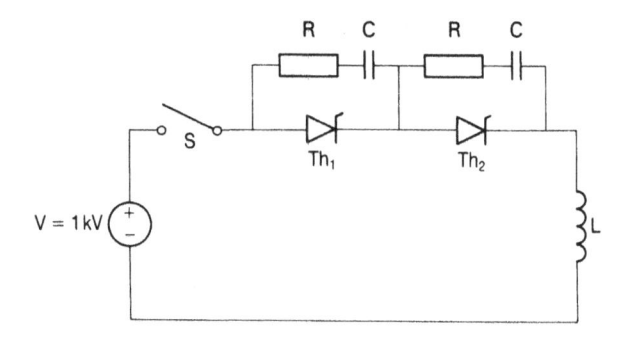

Figure 8.37

The circuit is shown in Fig. 8.37 and it should be noted that the thyristors Th_1 and Th_2 may be considered to be simply switches which are either closed (when they are turned on) or open (before they are turned on).

At the instant of closing the switch the current in the circuit is zero and the 1 kV is shared equally between the capacitors. When thyristor Th_1 turns on it short circuits the RC circuit in parallel with it and the circuit will then be as shown in Fig. 8.38, the R and the C being those in parallel with the still open switch Th_2. The transform circuit takes the form shown in Fig. 8.39.

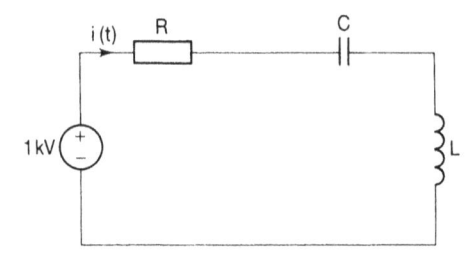

Figure 8.38

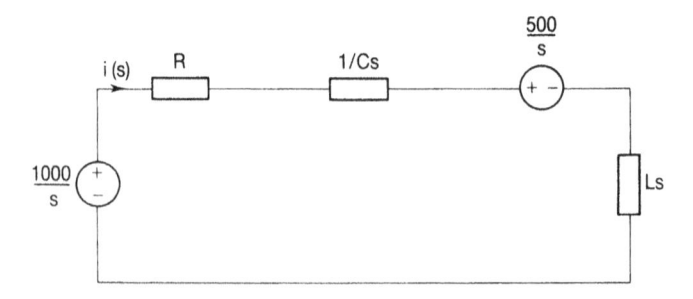

Figure 8.39

From the transform circuit we have that

$$i(s) = [(1000/s) - (500/s)]/(R + 1/Cs + Ls)$$
$$= 500/s(R + 1/Cs + Ls)$$
$$= 500/Ls^2 + Rs + 1/C$$
$$= (500/L)/[s^2 + (R/L)s + 1/CL]$$

Completing the square of the denominator we have

$$i(s) = (500/L)/[\{s + (R/2L)\}^2 + \{(1/CL) - (R/2L)^2\}]$$

Putting $L = 250\ \mu H$, $R/2L = \alpha$, and $\{(1/CL) - (R/2L)^2\} = \omega^2$, we have

$$i(s) = 2/[(s + \alpha)^2 + \omega^2]\ \mu s$$

Multiplying numerator and denominator by ω we have

$$i(s) = (2/\omega)\{\omega/[(s + \alpha)^2 + \omega^2]\}$$

From transform pair number 7 in Table 8.1 we see that $\omega/[(s + \alpha)^2 + \omega^2]$ is the transform of $\exp(-\alpha t)\sin \omega t$ so that, as a function of time we have for the current,

$$i(t) = (2/\omega)\exp(-\alpha t)\sin \omega t$$

This is of the form shown in Fig. 8.40 which is an exponentially decaying sine wave. It is said to be underdamped.

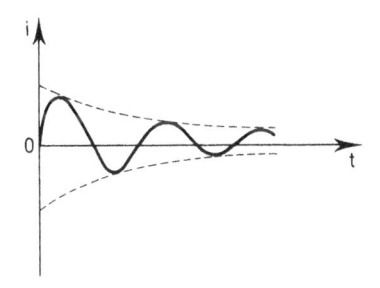

Figure 8.40

This means that the current undergoes a period of oscillation before reaching its new required value (zero in this case).

This result was obtained assuming that ω^2 is positive (i.e. $(1/CL) > (R/2L)^2$). There are other possibilities (ω^2 could be negative or zero) leading to other results associated with overdamping and critical damping, respectively, in which the current reaches zero more or less rapidly and without oscillation.

8.5 SELF-ASSESSMENT TEST

1 Give two conditions which could lead to the transient operation of an electric circuit.

2 Explain why transient conditions do not exist in purely resistive circuits.

3 State the reason why there could be periods of transient operation in inductive circuits.

4 Explain why there is a period of transient operation immediately after switching on a circuit containing capacitance.

5 Define a step voltage.

6 Give the symbol for and the unit of time constant.

7 Give an expression for the time constant of an RL circuit.

8 A coil has an inductance of 10 mH and a resistance of 5 Ω. If a step voltage of 10 V is applied to the coil, how long will it take for the current to reach its final steady state value?

9 What is the initial rate of rise of the current in the coil in Question 8?

10 What is the final steady state value of the current in the coil of Question 8?

11 State the form of the voltage across the capacitor in a series RC circuit to which a step voltage is applied.

12 State the form of the current in the circuit of Question 11.

13 A series RL circuit is to be used as an integrator. Across which circuit element should the output voltage be taken?

14 A series RC circuit is to be used as a differentiator. Across which circuit element should the output voltage be taken?

15 What is the value of the time constant of an RC series circuit having $R = 5$ kΩ and $C = 0.1$ μF?

16 A 10 μF capacitor, charged to 100 V is discharged through a resistance of 100 Ω. How long will it take to become fully discharged.

17 What will be the current in the circuit of Question 16 after a time equal to the time constant of the circuit?

18 An RC circuit having a time constant of 10 μs is connected to a 100 V d.c. supply. If it is partially charged to 50 V and then immediately discharged through the resistance, how long will it take to become fully discharged?

19 What is the advantage of using Laplace transforms in the analysis of the transient operation of electric circuits?

20 What is a transform circuit?

8.6 PROBLEMS

1 A step voltage of amplitude 100 V is applied to a coil of inductance 10 H and resistance 10 Ω. Determine (a) the initial rate of rise of current in the coil in ampere per second, (b) the value of the current after 0.1 s, (c) the circuit time constant, (d) the final steady state value of the current and (e) the time taken for the current to reach its steady state value.

2 The armature of a relay operates on a 200 V supply and is actuated when the current through its coil is 240 mA. The relay is required to close 4 ms after switching on the circuit. This time is also the time constant of the circuit. Calculate (a) the resistance and inductance of the coil and (b) the initial rate of rise of the current.

3 A series circuit consists of a resistor R ($=10\,\Omega$) in series with a coil having an inductance of 2 H and a resistance of 10 Ω. The circuit is fed from a 100 V d.c. supply and after it has been on for several minutes a short circuit fault suddenly occurs across the resistor R. Determine (a) the current in the circuit 0.1 s after the occurrence of the fault and (b) the time taken for the current to reach its new steady state value.

4 A resistor is connected in series with a 2 μF capacitor across a 200 V d.c. supply. A neon lamp with a striking voltage of 120 V is connected across the capacitor. Calculate the value of the resistance required to make the lamp flash 5 s after switching the circuit on.

5 An imperfect capacitor of 10 μF capacitance is fully charged from a 200 V d.c. supply. After being disconnected from the supply the voltage across its plates falls to 100 V in 20 s. Calculate the leakage resistance of the capacitor.

6 A 2 μF capacitor is connected in series with a 2 kΩ resistance to a 100 V d.c. supply having an effective internal resistance of 2 kΩ. The supply is switched on for 10 ms and then replaced by a short circuit. Determine the total time for which the capacitor voltage is greater than 10 V. Sketch the waveform of the voltage across the capacitor and of the current through it.

7 From the instant that the switch S in the circuit of Fig. 8.41 is closed, (a) show that the current through the capacitor is given by
$i_C = (V/R_1) \exp[-R_1 + R_2)t/CR_1R_2]$, (b) obtain numerical expressions for the voltage across the capacitor and the current supplied by the source.

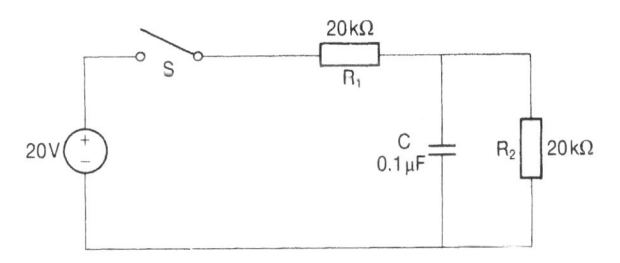

Figure 8.41

8 The RL integrator circuit shown in Fig. 8.9 has $R = 47\,\Omega$ and $L = 120$ mH. Obtain the output voltage (v_o) waveform when the input voltage V_{in} is:
(a) a step function of amplitude 20 V;

(b) a single pulse of amplitude 20 V and width 5 ms;
(c) a train of pulses of amplitude 20 V, width 5 ms and separation 20 ms;

9 The RC differentiator circuit shown in Fig. 8.26 has $R = 1.5$ MΩ and $C = 2$ pF. Obtain the output voltage (v_o) waveform when the input voltage (v_{in}) is:
(a) a step function of amplitude 15 V;
(b) a single pulse of amplitude 15 V and width 5 μs;
(c) a train of pulses of amplitude 15 V, width 5 μs and separation 20 μs;
(d) a train of pulses of amplitude 15 V, width 5 μs and separation 5 μs.

10 A series circuit consists of a 50 Ω resistor, a 0.1 μF capacitor and a 250 μH inductor in series. With the capacitor initially uncharged, the circuit is connected to a 500 V d.c. supply. Obtain an expression for the current in the circuit as a function of time.

9 Two-port networks

9.1 INTRODUCTION

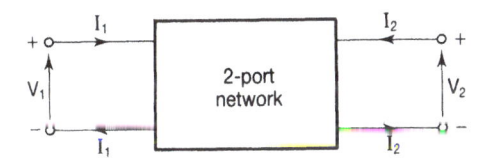

Figure 9.1

A two-port network has four terminals as shown in Fig. 9.1 and is often called a four-terminal network. These are made up of:

1. an input pair by which an input current (I_1 say) enters one terminal and leaves the other; and

2. an output pair by which an output current (I_2 say) enters one terminal and leaves the other.

The input terminals constitute the input port and the output terminals constitute the output port. The conventional directions of the currents are as shown in the diagram.

Some examples of two-port networks are: transistor circuits; amplifier circuits; filters, power transmission lines.

For linear, passive networks a set of equations can be established which relate input and output quantities in terms of the network impedances or admittances. If the input and output currents and voltages (I_1, I_2, V_1 and V_2) are considered in pairs to be the independent variables, the others then being the dependent variables, there are six possible pairs or sets of equations. These will be discussed in turn.

9.2 THE IMPEDANCE OR z-PARAMETERS

Set 1: I_1 and I_2 are the independent variables. The dependent variables are then given by

$$V_1 = z_{11}I_1 + z_{12}I_2 \tag{9.1}$$

$$V_2 = z_{21}I_1 + z_{22}I_2 \tag{9.2}$$

In matrix form these may be written

$$\begin{bmatrix} V_1 \\ V_2 \end{bmatrix} = \begin{bmatrix} z_{11} & z_{12} \\ z_{21} & z_{22} \end{bmatrix} \begin{bmatrix} I_1 \\ I_2 \end{bmatrix} \tag{9.3}$$

The parameters for this set are called the impedance or z-parameters and they are defined as follows:

- z_{11} is the input impedance (z_i) and is measured as V_1/I_1 with $I_2 = 0$:

$$z_{11} = (V_1/I_1)|_{I_2=0} \tag{9.4}$$

- z_{12} is the forward transfer impedance (z_f) and is measured as V_1/I_2 with $I_1 = 0$:

$$z_{12} = (V_1/V_2)|_{I_1=0} \tag{9.5}$$

- z_{21} is the reverse transfer impedance (z_r) and is measured as V_2/I_1 with $I_2 = 0$:

$$z_{21} = (V_2/I_1)|_{I_2=0} \tag{9.6}$$

- z_{22} is the output impedance (z_o) and is measured as V_2/I_2 with $I_1 = 0$:

$$z_{22} = (V_2/I_2)|_{I_1=0} \tag{9.7}$$

Since these are all obtained with either the input terminals or the output terminals open circuited, they are called the open circuit impedance parameters. They are measured in ohms.

The circuit shown in Fig. 9.2 satisfies Equations (9.1) and (9.2) and is called the equivalent circuit for this set.

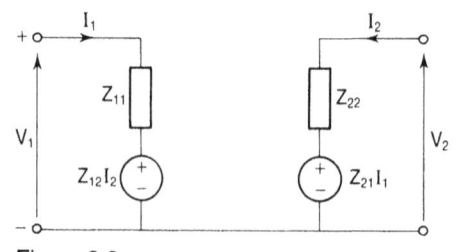

Figure 9.2

Example 9.1

Determine the z-parameters of the T-circuit shown in Fig. 9.3.

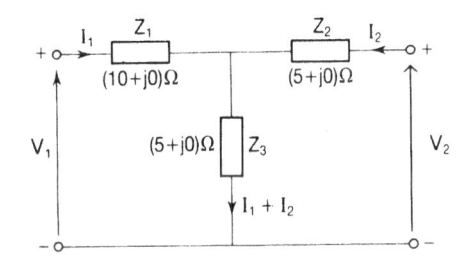

Figure 9.3

Solution

From Equation (9.4) $z_{11} = V_1/I_1$ with $I_2 = 0$ and applying KVL with $I_2 = 0$ we have

$$V_1 = I_1 Z_1 + I_1 Z_3 = I_1(Z_1 + Z_3)$$

so

$$z_{11} = V_1/I_1 = Z_1 + Z_3$$

From Equation (9.6) $z_{21} = V_2/I_1$ with $I_2 = 0$ and with $I_2 = 0$, $V_2 = I_1 Z_3$, so

$$z_{21} = V_2/I_1 = Z_3$$

From Equation (9.7) $z_{22} = V_2/I_2$ with $I_1 = 0$ and applying KVL with $I_1 = 0$ we have

$$V_2 = I_2 Z_2 + I_2 Z_3 = I_2(Z_2 + Z_3).$$

Therefore

$$z_{22} = V_2/I_2 = Z_2 + Z_3$$

From Equation (9.5) $z_{12} = V_1/I_2$ with $I_1 = 0$ and with $I_1 = 0$, $V_1 = I_2 Z_3$, so

$$z_{12} = V_1/I_2 = Z_3$$

Note that $z_{12} = z_{21}$.

Putting in the values for Z_1, Z_2 and Z_3 from the diagram we have

$$z_{11} = Z_1 + Z_3 = 10 + 5 = 15 \, \Omega$$

$$z_{22} = Z_2 + Z_3 = 5 + 5 = 10 \, \Omega$$

$$z_{12} = z_{21} = Z_3 = 5 \, \Omega$$

In matrix form we have

$$\begin{bmatrix} V_1 \\ V_2 \end{bmatrix} = \begin{bmatrix} 15 & 5 \\ 5 & 10 \end{bmatrix} \begin{bmatrix} I_1 \\ I_2 \end{bmatrix}$$

9.3 THE ADMITTANCE OR y-PARAMETERS

Set 2: If V_1 and V_2 are the independent variables then the dependent variables are given by:

$$I_1 = y_{11}V_1 + y_{12}V_2 \tag{9.8}$$

$$I_2 = y_{21}V_1 + y_{22}V_2 \tag{9.9}$$

In matrix form

$$\begin{bmatrix} I_1 \\ I_2 \end{bmatrix} = \begin{bmatrix} y_{11} & y_{12} \\ y_{21} & y_{22} \end{bmatrix} \begin{bmatrix} V_1 \\ V_2 \end{bmatrix} \tag{9.10}$$

The parameters for this set are called the admittance or y-parameters and they are defined as follows:

- y_{11} is the input admittance (y_i) and is measured as I_1/V_1 with $V_2 = 0$:

$$y_{11} = (I_1/V_1)|_{V_2=0} \tag{9.11}$$

- y_{12} is the reverse transfer admittance (y_r) and is measured as I_1/V_2 with $V_1 = 0$:

$$y_{12} = (I_1/V_2)|_{V_1=0} \tag{9.12}$$

- y_{21} is the forward transfer admittance (y_f) and is measured as I_2/V_1 with $V_2 = 0$:

$$y_{21} = (I_2/V_1)|_{V_2=0} \tag{9.13}$$

- y_{22} is the output admittance (y_o) and is measured as I_2/V_2 with $V_1 = 0$:

$$y_{22} = (I_2/V_2)|_{V_1=0} \tag{9.14}$$

Since these are obtained with the output or the input port short circuited, they are called the short circuit admittance parameters and are measured in siemens. They are commonly used in the analysis of field effect transistor (FET) circuits.

The equivalent circuit for this set which satisfies Equations (9.8) and (9.9) is given in Fig. 9.4.

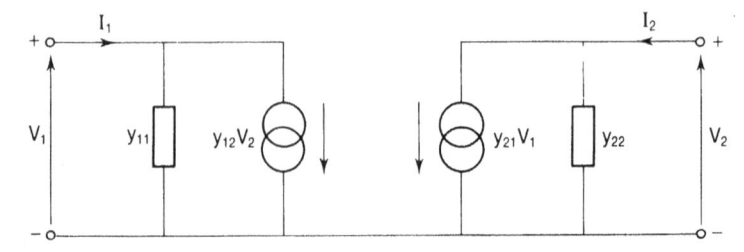

Figure 9.4

Example 9.2

Determine the admittance parameters for the T-circuit shown in Fig. 9.3.

Solution

From Equation (9.11), $y_{11} = I_1/V_1$ with $V_2 = 0$. Short circuiting the output port terminals makes $V_2 = 0$ and then Z_2 and Z_3 are in parallel. Now

$$V_1 = I_1[Z_1 + \{Z_2 Z_3/(Z_2 + Z_3)\}]$$

so that

$$I_1 = V_1/[Z_1 + \{Z_2 Z_3/(Z_2 + Z_3)\}]$$

Putting in the impedance values we have

$$I_1 = V_1/(10 + 2.5) = V_1/12.5$$

Therefore

$$y_{11} = I_1/V_1 = (1/12.5) \text{ S}$$

From Equation (9.12), $y_{12} = I_1/V_2$ with $V_1 = 0$. Short circuiting the input terminals to make $V_1 = 0$ places Z_1 and Z_3 in parallel. Now

$$V_2 = I_2[Z_2 + \{Z_1 Z_3/(Z_1 + Z_3)\}]$$

so that

$$I_2 = V_2/[Z_2 + \{Z_1 Z_3/(Z_1 + Z_3)\}]$$

Putting in the impedance values we have

$$I_2 = V_2/[5 + (50/15)] = V_2/[(75 + 50)/15] = 15 V_2/125$$

By current division

$$I_1 = -I_2[Z_3/(Z_1 + Z_3)] = -(15 V_2/125)(5/15) = -V_2/25$$

so

$$y_{12} = I_1/V_2 = (-1/25) \text{ S}$$

From Equation (9.13), $y_{21} = I_2/V_1$ with $V_2 = 0$. We have seen above that, with the output port short circuited to make $V_2 = 0$, $I_1 = V_1/12.5$. By current division,

$$I_2 = -I_1[Z_3/(Z_2 + Z_3)] = -(V_1/12.5)(5/10) = (-V_1/25)$$

Therefore

$$y_{21} = I_2/V_1 = (-1/25) \text{ S}.$$

Note that $y_{12} = y_{21}$.

210 Two-port networks

Finally, from Equation (9.14), $y_{22} = I_2/V_2$ with $V_1 = 0$. We have seen that with $V_1 = 0$,

$$I_2 = 15V_2/125 = 3V_2/25$$

so

$$y_{22} = I_2/V_2 = (3/25) \text{ S}$$

In matrix form we have

$$\begin{bmatrix} I_1 \\ I_2 \end{bmatrix} = \begin{bmatrix} 1/12.5 & -1/25 \\ -1/25 & 3/25 \end{bmatrix} \begin{bmatrix} V_1 \\ V_2 \end{bmatrix}$$

9.4 THE HYBRID OR h-PARAMETERS

Set 3: If I_1 and V_2 are the independent variables then the dependent variables are given by:

$$V_1 = h_{11}I_1 + h_{12}V_2 \tag{9.15}$$

$$I_2 = h_{21}I_1 + h_{22}V_2 \tag{9.16}$$

In matrix form

$$\begin{bmatrix} V_1 \\ I_1 \end{bmatrix} = \begin{bmatrix} h_{11} & h_{12} \\ h_{21} & h_{22} \end{bmatrix} \begin{bmatrix} I_1 \\ V_2 \end{bmatrix} \tag{9.17}$$

The parameters of this set are called the hybrid or h-parameters and they are defined as follows:

- h_{11} is the input impedance (h_i) and is measured in ohms as V_1/I_1 with $V_2 = 0$:

$$h_{11} = (V_1/I_1)|_{V_2=0} \tag{9.18}$$

- h_{21} is the forward current gain (h_f) and is a dimensionless ratio of currents (I_2/I_1) with $V_2 = 0$:

$$h_{21} = (I_2/I_1)|_{V_2=0} \tag{9.19}$$

- h_{12} is the reverse voltage gain (h_r) and is a dimensionless ratio of voltages (V_1/V_2) with $I_1 = 0$:

$$h_{12} = (V_1/V_2)|_{I_1=0} \tag{9.20}$$

- h_{22} is the output admittance (h_o) and is measured in siemens as I_2/V_2 with $I_1 = 0$

$$h_{22} = (I_2/V_2)|_{I_1=0} \tag{9.21}$$

These parameters are used extensively in the small signal analysis of bipolar transistors where an additional subscript 'e', 'b' or 'c' is used, depending upon the configuration (common emitter, common base or common collector, respectively). Thus we might have for the input impedance h_{1e} or h_{1b} or h_{1c}.

The Equations (9.15) and (9.16) are satisfied by the equivalent circuit shown in Fig. 9.5.

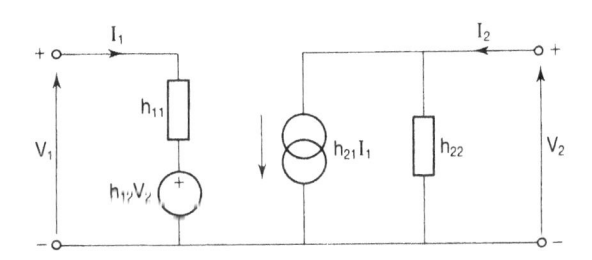

Figure 9.5

9.5 THE INVERSE HYBRID OR g-PARAMETERS

Set 4: If V_1 and I_2 are the independent variables then the dependent variables are given by:

$$I_1 = g_{11}V_1 - g_{12}I_2 \tag{9.22}$$

$$V_2 = g_{21}V_1 + g_{22}I_2 \tag{9.23}$$

In matrix form

$$\begin{bmatrix} I_1 \\ V_2 \end{bmatrix} = \begin{bmatrix} g_{11} & g_{12} \\ g_{21} & g_{22} \end{bmatrix} \begin{bmatrix} V_1 \\ I_2 \end{bmatrix} \tag{9.24}$$

The parameters of this set are called the inverse hybrid or g-parameters and are defined as follows:

- g_{11} is the input admittance (g_i) and is measured in siemens as I_1/V_1 with $I_2 = 0$:

$$g_{11} = (I_1/V_1)|_{I_2=0} \tag{9.25}$$

- g_{21} is the forward voltage gain (g_f) and is a dimensionless ratio of voltages (V_2/V_1) with $I_2 = 0$:

$$g_{21} = (V_2/V_1)|_{I_2=0} \tag{9.26}$$

- g_{12} is the reverse current gain (g_r) and is a dimensionless ratio of currents (I_1/I_2) with $V_1 = 0$:

$$g_{12} = (I_1/I_2)|_{V_1=0} \tag{9.27}$$

212 Two-port networks

- g_{22} is the output impedance (g_o) and is measured in ohms as V_2/I_2 with $V_1 = 0$:

$$g_{22} = (V_2/I_2)|_{V_1=0} \tag{9.28}$$

Note that

$$g_{11} = 1/h_{11} \tag{9.29}$$

$$g_{21} = 1/h_{21} \tag{9.30}$$

$$g_{12} = 1/h_{12} \tag{9.31}$$

$$g_{22} = 1/h_{22} \tag{9.32}$$

For this reason the g-parameters are also known as the inverse h-parameters. Equations (9.22) and (9.23) are satisfied by the equivalent circuit of Fig. 9.6.

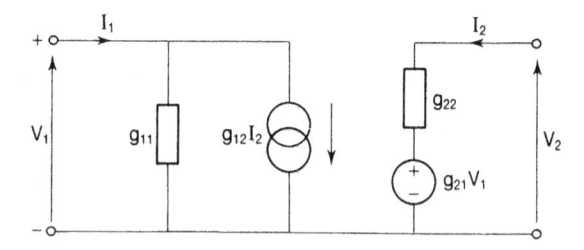

Figure 9.6

Example 9.3

Determine the h- and g-parameters for the T-circuit of Fig. 9.3.

Solution

From Equation (9.18), $h_{11} = V_1/I_1$ with $V_2 = 0$. Short circuiting the output port terminals to make $V_2 = 0$ places Z_2 and Z_3 in parallel and then

$$I_1 = V_1/[Z_1 + \{Z_2Z_3/(Z_2 + Z_3)\}]$$

Therefore

$$h_{11} = V_1/I_1 = Z_1 + \{Z_2Z_3/(Z_2 + Z_3)\}$$

Putting in the impedance values we have

$$h_{11} = (10 + (5 \times 5)/10) = 12.5 \, \Omega$$

From Equation (9.29)

$$g_{11} = 1/h_{11} = (1/12.5) \text{ S}$$

From Equation (9.19), $h_{21} = I_2/I_1$ with $V_2 = 0$. By current division

$$I_2 = -I_1[Z_3/(Z_2 + Z_3)]$$

Therefore

$$h_{21} = I_2/I_1 = -Z_3/(Z_2 + Z_3)$$

Putting in the impedance values,

$$h_{21} = -5/10 = -0.5$$

From Equation (9.30)

$$g_{21} = 1/h_{21} = -2$$

From Equation (9.21), $h_{22} = I_2/V_2$ with $I_1 = 0$. With $I_1 = 0$, $I_2 = V_2/(Z_2 + Z_3)$

so

$$h_{22} = I_2/V_2 = 1/(Z_2 + Z_3) \text{ S}$$

Putting in the impedance values

$$h_{22} = 1/(5 + 5) = 0.1 \text{ S}$$

From Equation (9.32)

$$g_{22} = 1/h_{22} = (Z_2 + Z_3) = (5 + 5) = 10 \text{ }\Omega$$

From Equation (9.20), $h_{12} = V_1/V_2$ with $I_1 = 0$. Now with the input port open circuited to make $I_1 = 0$, $V_1 = I_2 Z_3$ and $I_2 = V_2/(Z_2 + Z_3)$,

$$V_1 = V_2 Z_3/(Z_2 + Z_3)$$

It follows that

$$h_{12} = V_1/V_2 = Z_3/(Z_2 + Z_3)$$

Putting in the impedance values we have

$$h_{12} = 5/(5 + 5) = 5/10 = 0.5$$

From Equation (9.31)

$$g_{12} = 1/h_{12} = 1/0.5 = 2$$

9.6 THE TRANSMISSION OR *ABCD*-PARAMETERS

Set 5: The fifth set, which leads to the so-called transmission or *ABCD*-parameters, has V_2 and I_2 as the independent variables and V_1, I_1 as the dependent variables. The output port current flows out of the positive terminal which is the opposite direction to the conventional direction considered in the previous four sets of equations. This is shown in Fig. 9.7.

These equations are used extensively for the analysis of transmission systems

214 Two-port networks

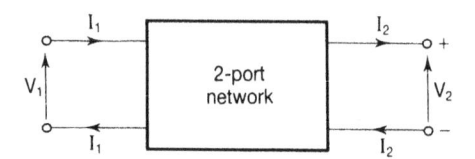

Figure 9.7

and are also particularly useful for the analysis of cascaded two-port networks, such as are frequently encountered in power systems.

The dependent variables are given by

$$V_1 = AV_2 + BI_2 \tag{9.33}$$

$$I_1 = CV_2 + DI_2 \tag{9.34}$$

In matrix form

$$\begin{bmatrix} V_1 \\ I_1 \end{bmatrix} = \begin{bmatrix} A & B \\ C & D \end{bmatrix} \begin{bmatrix} V_2 \\ I_2 \end{bmatrix} \tag{9.35}$$

The $ABCD$ matrix is called the transfer matrix of the network. The parameters for this set are measured by open circuit and short circuit tests on the output port and are defined as follows:

- A is measured as V_1/V_2 with $I_2 = 0$ and, being the ratio of two voltages, it is dimensionless:

$$A = (V_1/V_2)|_{I_2=0} \tag{9.36}$$

- B is measured as V_1/I_2 with $V_2 = 0$ and, being a ratio of volts to amperes, has the dimensions of impedance:

$$B = (V_1/I_2)|_{V_2=0} \tag{9.37}$$

- C is measured as I_1/V_2 with $I_2 = 0$ and, being the ratio of amperes to volts, it has the dimensions of admittance:

$$C = (I_1/V_2)|_{I_2=0} \tag{9.38}$$

- D is measured as I_1/I_2 with $V_2 = 0$ and, being the ratio of two currents, it is dimensionless:

$$D = (I_1/I_2)|_{V_2=0} \tag{9.39}$$

9.7 THE INVERSE TRANSMISSION PARAMETERS

Set 6: If V_1 and I_1 are the independent variables, then the dependent variables are given by

$$V_2 = A'V_1 + B'I_1 \tag{9.40}$$

$$I_2 = C'V_1 + D'I_1 \tag{9.41}$$

In matrix form we have

$$\begin{bmatrix} V_2 \\ I_2 \end{bmatrix} = \begin{bmatrix} A' & B' \\ C' & D' \end{bmatrix} \begin{bmatrix} V_1 \\ I_1 \end{bmatrix} \tag{9.42}$$

A', B', C' and D' are called the inverse transmission parameters or the inverse ABCD-parameters and they are measured with the input port either open circuited or short circuited. They are defined as follows:

- A' is measured as V_2/V_1 with $I_1 = 0$ and is a dimensionless ratio of two voltages:

$$A' = (V_2/V_1)|_{I_1=0} \tag{9.43}$$

- B' is measured as V_2/I_1 with $V_1 = 0$ and its unit is the volt per ampere. It therefore has the dimensions of impedance:

$$B' = (V_2/I_1)|_{V_1=0} \tag{9.44}$$

- C' is measured as I_2/V_1 with $I_1 = 0$ and its unit is the ampere per volt. It therefore has the dimensions of admittance:

$$C' = (I_2/V_1)|_{I_1=0} \tag{9.45}$$

- D' is measured as I_2/I_1 with $V_1 = 0$ and is a dimensionless ratio of two currents:

$$D' = (I_2/I_1)|_{V_1=0} \tag{9.46}$$

Example 9.4

Determine the ABCD-parameters of the series impedance network shown in Fig. 9.8.

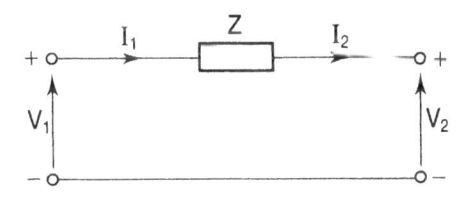

Figure 9.8

Solution

To find A and C we open circuit the output port to make $I_2 = 0$. Then $V_1 = V_2$ and $I_1 = I_2 = 0$. From Equation (9.36), $A = V_1/V_2 = 1$. From Equation (9.38), $C = I_1/V_2 = 0$.

216 Two-port networks

To find B and D we short circuit the output port to make $V_2 = 0$. Then $V_1 = I_2 Z$ and $I_1 = I_2$. From Equation (9.37), $B = V_1/I_2 = Z$. From Equation (9.39), $D = I_1/I_2 = 1$.

We could also find the *ABCD*-parameters by using Kirchhoff's laws as follows. Applying KVL to the network and taking the clockwise direction to be positive, we see that $V_1 - I_2 Z - V_2 = 0$. Rearranging we get

$$V_1 = V_2 + ZI_2 \tag{9.47}$$

Comparing Equations (9.33) and (9.47) we see that $A = 1$ and that $B = Z$.

Applying KCL to the circuit we see that $I_1 = I_2$, which may be written as

$$I_1 = 0V_2 + I_2 \tag{9.48}$$

Comparing Equations (9.34) and (9.48) we see that $C = 0$ and that $D = 1$.

In matrix form

$$\begin{bmatrix} V_1 \\ I_1 \end{bmatrix} = \begin{bmatrix} 1 & Z \\ 0 & 1 \end{bmatrix} \begin{bmatrix} V_2 \\ I_2 \end{bmatrix} \tag{9.49}$$

Note that $A = D$. This is always the case for symmetrical two-port networks, that is networks for which the input and output ports are interchangeable. A two-port network which is not symmetrical is shown in Example 9.6.

'Short' power transmission lines

A power transmission line will have conductor resistance and inductance distributed along the length of the line and capacitance between conductors also distributed along the length of the line. A 'short' line (less than about 80 km in length) is one for which, for the purposes of analysis, the capacitance can be neglected and the resistance and inductance can be considered to be concentrated at the centre of the line without introducing great errors. The line may then be represented by the series impedance network of Fig. 9.8.

Example 9.5

A single-phase transmission line has an impedance $Z = (0.22 + j0.36)$ Ω. (1) determine the *ABCD*-parameters of this line, and (2) calculate the sending-end voltage required to produce 500 kVA at a voltage of 2 kV when operating at unity power factor.

Solution

The network is as shown in Fig. 9.8. From the matrix Equation (9.49) we see that $A = D = 1$; $B = Z = (0.22 + j0.36)$ Ω; $C = 0$. Taking the receiving-end voltage as the reference,

$$V_2 = (2000 + j0) \text{ V}$$

The receiving-end current is given by

$I_2 = 500 \text{ kVA}/2 \text{ kV} = 250 \text{ A}$

Because the receiving-end power factor is unity, I_2 is in phase with V_2 and so

$I_2 = (250 + j0) \text{ A}$

From Equation (9.47)

$V_1 = V_2 + ZI_2$

Therefore

$V_1 = (2000 + j0) + (250 + j0)(0.22 + j0.36)$
$= 2000 + 55 + j90$
$= 2055 + j90$
$= \sqrt{(2055^2 + 90^2)} \angle \tan^{-1}(90/2055)$
$= 2057 \angle 2.51° \text{ V}$

Example 9.6

Determine the $ABCD$-parameters for the network shown in Fig. 9.9.

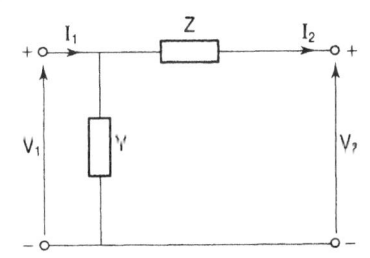

Figure 9.9

Solution

With the output port open circuited to make $I_2 = 0$, $I_1 = V_1 Y$ and $V_2 = V_1$. From Equation (9.36), $A = V_1/V_2 = 1$. From Equation (9.38), $C = I_1/V_2 = I_1/V_1 = Y$.

With the output port short circuited to make $V_2 = 0$, we see, using the current division technique, that

$I_2 = I_1[(1/Y)/\{(1/Y) + Z\}]$
$= I_1[(1/Y)/\{(1 + ZY)/Y\}]$
$= I_1/(1 + ZY)$

From Equation (9.39), $D = I_1/I_2$, so

$D = 1 + ZY$

From Equation (9.37), $B = V_1/I_2$. Now

$I_1 = V_1/[\{(1/Y)(Z)/((1/Y) + Z)\}] = V_1[(1 + ZY)/Y]/(Z/Y) = V_1(1 + ZY)/Z$

218 *Two-port networks*

Therefore

$$I_1/(1 + ZY) = V_1/Z$$

Also, we saw above that $I_2 = I_1/(1 + ZY)$ so that $I_2 = V_1/Z$. Therefore

$$B = V_1/I_2 = Z$$

In matrix form

$$\begin{bmatrix} V_1 \\ I_1 \end{bmatrix} = \begin{bmatrix} 1 & Z \\ Y & 1 + ZY \end{bmatrix} \begin{bmatrix} V_2 \\ I_2 \end{bmatrix} \quad (9.50)$$

Note that in this case $A \neq D$. This two-port network is not symmetrical because the input and output ports are not interchangeable.

Example 9.7

Obtain the *ABCD*-parameters for the network shown in Fig. 9.10.

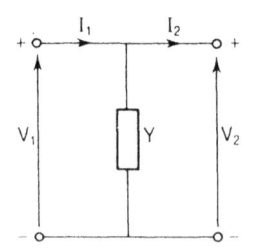

Figure 9.10

Solution

This is called a shunt admittance network. From Equation (9.36), $A = V_1/V_2|_{I_2=0}$. With $I_2 = 0$, $V_1 = V_2$ so that $A = 1$.

From Equation (9.38), $C = I_1/V_2|_{I_2=0}$. With $I_2 = 0$, $I_1 = V_1 Y = V_2 Y$ since $V_1 = V_2$. Therefore

$$C = I_1/V_2 = Y$$

From Equation (9.39), $D = I_1/I_2|_{V_2=0}$. With $V_2 = 0$, $I_1 = I_2$ since Y is short circuited, so that $D = 1$. As expected then for this symmetrical network, $A = D$.

From Equation (9.37), $B = V_1/I_2|_{V_2=0}$. With $V_2 = 0$, $V_1 = 0$ and so $B = 0$.

In matrix form

$$\begin{bmatrix} V_1 \\ I_1 \end{bmatrix} = \begin{bmatrix} 1 & 0 \\ Y & 1 \end{bmatrix} \begin{bmatrix} V_2 \\ I_2 \end{bmatrix} \quad (9.51)$$

9.8 CASCADED TWO-PORT NETWORKS

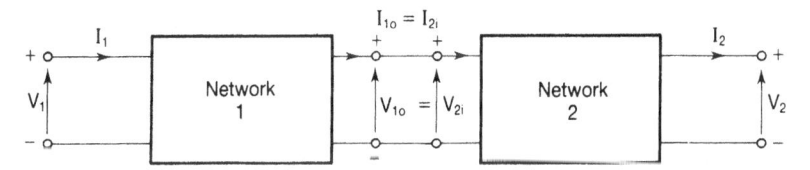

Figure 9.11

Two networks connected as shown in Fig. 9.11 are said to be cascaded. The output port of network 1 is the input port of network 2. The *ABCD*-parameters of the cascaded pair are obtained by the multiplication of those of the two constituent networks. Thus, if the transmission parameters of circuit 1 are A_1, B_1, C_1 and D_1, while those for network 2 are A_2, B_2, C_2 and D_2 then

$$\begin{bmatrix} V_1 \\ I_1 \end{bmatrix} = \begin{bmatrix} A_1 & B_1 \\ C_1 & D_1 \end{bmatrix} \begin{bmatrix} A_2 & B_2 \\ C_2 & D_2 \end{bmatrix} \begin{bmatrix} V_2 \\ I_2 \end{bmatrix} = \begin{bmatrix} A & B \\ C & D \end{bmatrix} \begin{bmatrix} V_2 \\ I_2 \end{bmatrix} \quad (9.52)$$

where, by matrix multiplication,

$$A = (A_1 A_2 + B_1 C_2) \quad (9.53)$$
$$B = (A_1 B_2 + B_1 D_2) \quad (9.54)$$
$$C = (C_1 A_2 + D_1 C_2) \quad (9.55)$$
$$D = (C_1 B_2 + D_1 D_2) \quad (9.56)$$

Example 9.8

Obtain the *ABCD*-parameters for the network of Fig. 9.9.

Solution

The two-port network of Fig. 9.9 may be considered to be a cascaded pair comprising a shunt admittance followed by a series impedance as shown in Fig. 9.12.

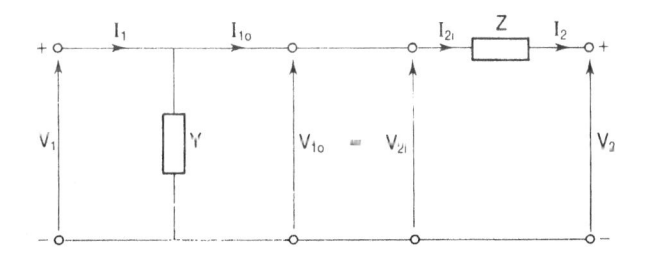

Figure 9.12

We saw in Example 9.7 that for a shunt admittance $A = 1$, $B = 0$, $C = Y$ and $D = 1$. Also, from Example 9.4 we have, for series impedance, $A = 1$, $B = Z$, $C = 0$ and $D = 1$.

Using the matrix Equation (9.52), with $A_1 = 1$, $B_1 = 0$, $C_1 = Y$, $D_1 = 1$, $A_2 = 1$, $B_2 = Z$, $C_2 = 0$ and $D_2 = 1$ we have

$$\begin{bmatrix} V_1 \\ I_1 \end{bmatrix} = \begin{bmatrix} 1 & 0 \\ Y & 1 \end{bmatrix} \begin{bmatrix} 1 & Z \\ 0 & 1 \end{bmatrix} \begin{bmatrix} V_2 \\ I_2 \end{bmatrix} = \begin{bmatrix} A & B \\ C & D \end{bmatrix} \begin{bmatrix} V_2 \\ I_2 \end{bmatrix}$$

From Equations (9.53)–(9.56) by matrix multiplication we have

$A = [(1 \times 1) + (0 \times 0)] = 1$
$B = [(1 \times Z) + (0 \times 1)] = Z$
$C = [(Y \times 1) + (1 \times 0)] = Y$
$D = [(Y \times Z) + (1 \times 1)] = 1 + ZY$

For the cascaded network, therefore, $A = 1$, $B = Z$, $C = Y$ and $D = 1 + ZY$. These results agree with those obtained in Example 9.6.

The *ABCD*-parameters of a π-network

The π- and the T-networks are commonly encountered in electric circuit theory, for example in filter circuits, attenuator sections and power transmission circuits. The π-network is essentially a delta network and a T-network is a star connection. As we saw in Chapter 3 it is possible to transform from one to the other using the delta-star transform or the star-delta transform. The more appropriate form for any given circuit application can thus be chosen.

Example 9.9

Obtain the transmission parameters for the π-network shown in Fig. 9.13.

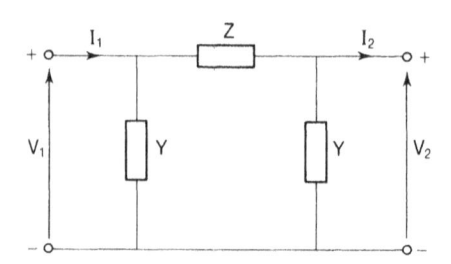

Figure 9.13

Solution

This circuit is made up from three two-port networks connected in cascade as indicated in Fig. 9.14.

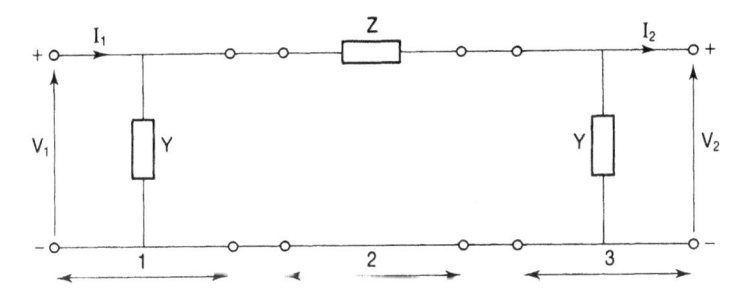

Figure 9.14

For sections 1 and 3, which are shunt admittances, $A = 1$, $B = 0$, $C = Y$ and $D = 1$. For the series impedance section, $A = 1$, $B = Z$, $C = 0$ and $D = 1$. Thus for the π-network

$$\begin{bmatrix} V_1 \\ I_1 \end{bmatrix} = \begin{bmatrix} 1 & 0 \\ Y & 1 \end{bmatrix} \begin{bmatrix} 1 & Z \\ 0 & 1 \end{bmatrix} \begin{bmatrix} 1 & 0 \\ Y & 1 \end{bmatrix} \begin{bmatrix} V_2 \\ I_2 \end{bmatrix}$$

Multiplying the first two matrices on the right-hand side of the equation we get

$$\begin{bmatrix} V_1 \\ I_1 \end{bmatrix} = \begin{bmatrix} 1 & Z \\ Y & 1+ZY \end{bmatrix} \begin{bmatrix} 1 & 0 \\ Y & 1 \end{bmatrix} \begin{bmatrix} V_2 \\ I_2 \end{bmatrix}$$

Finally, multiplying the remaining transfer matrices, we have

$$\begin{bmatrix} V_1 \\ I_1 \end{bmatrix} = \begin{bmatrix} 1+ZY & Z \\ Y+(1+ZY)Y & 1+ZY \end{bmatrix} \begin{bmatrix} V_2 \\ I_2 \end{bmatrix} \qquad (9.57)$$

For the π-network, therefore, $A = 1 + ZY$, $B = Z$, $C = Y + (1 + ZY)Y = 2Y + ZY^2$, $D = 1 + ZY$. Again $A = D$, this being a symmetrical two-port network.

Nominal-π representation of 'medium length' power transmission lines

In power transmission the π-circuit of Fig. 9.13 is referred to as a nominal-π network. It is used to model a medium length transmission line (between 80 km and 200 km). The whole of the impedance of the line is assumed to be concentrated at the centre of the line and half the capacitive reactance is placed at either end of the line. Thus, if we replace Y by $Y/2$, where Y is the total shunt admittance of the line, we obtain for the $ABCD$-parameters:

$$A = D = 1 + (ZY/2), \quad B = Z, \quad C = Y + (ZY^2/4)$$

The *ABCD*-parameters of a T-network

T-networks are commonly used for LC filters and for attenuation sections.

Example 9.10

Obtain the *ABCD*-parameters for the T-network shown in Fig. 9.15.

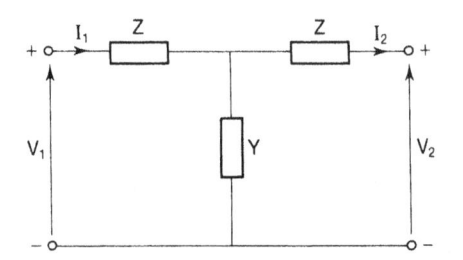

Figure 9.15

Solution

This network may be considered to be made up of a series impedance network followed by a shunt admittance network and then another series impedance network all connected in cascade. We can then make use of the transfer matrices in Equations (9.49) and (9.51) to write down the matrix equation for this circuit:

$$\begin{bmatrix} V_1 \\ I_1 \end{bmatrix} = \begin{bmatrix} 1 & Z \\ 0 & 1 \end{bmatrix} \begin{bmatrix} 1 & 0 \\ Y & 1 \end{bmatrix} \begin{bmatrix} 1 & Z \\ 0 & 1 \end{bmatrix} \begin{bmatrix} V_2 \\ I_2 \end{bmatrix}$$

By multiplication of the first two transfer matrices we get

$$\begin{bmatrix} V_1 \\ I_1 \end{bmatrix} = \begin{bmatrix} (1+ZY) & (0+Z) \\ (0+Y) & (0+1) \end{bmatrix} \begin{bmatrix} 1 & Z \\ 0 & 1 \end{bmatrix} \begin{bmatrix} V_2 \\ I_2 \end{bmatrix}$$

Multiplying the remaining two transfer matrices we have

$$\begin{bmatrix} V_1 \\ I_1 \end{bmatrix} = \begin{bmatrix} (1+ZY+0) & (Z+Z^2Y+Z) \\ (Y+0) & (ZY+1) \end{bmatrix} \begin{bmatrix} V_2 \\ I_2 \end{bmatrix}$$

Finally

$$\begin{bmatrix} V_1 \\ I_1 \end{bmatrix} = \begin{bmatrix} 1+ZY & 2Z+Z^2Y \\ Y & 1+ZY \end{bmatrix} \begin{bmatrix} V_2 \\ I_2 \end{bmatrix} \quad (9.58)$$

The transmission parameters for this network are therefore

$A = D = 1 + ZY; \quad B = 2Z + Z^2Y; \quad C = Y$

Nominal-T representation of 'medium length' power transmission lines

In power transmission lines this T-network is called a nominal-T network. Half of the line impedance is considered to be concentrated at each end of the line and the whole of the shunt admittance is placed at the centre of the line. Replacing the impedance Z by $Z/2$, where Z is the total series impedance of the line, gives for the transmission parameters of a nominal-T network:

$$A = D = 1 + ZY/2; \quad B = Z + Z^2Y/4; \quad C = Y$$

The *ABCD*-parameters of a single-phase, two-winding transformer

The transformer is a very important electrical 'machine', used not only for voltage and current level changing but also for buffering and matching purposes. They are manufactured in an enormous range of sizes from a few VA in electronic circuits to more than 1000 MVA in power systems.

Example 9.11

Determine the *ABCD*-parameters for the ideal two-winding transformer shown in Fig. 9.16.

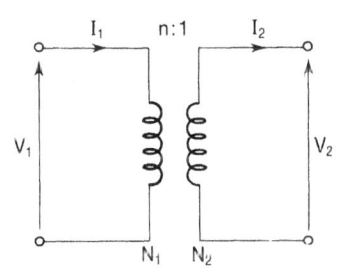

Figure 9.16

Solution

The ideal transformer shown has a transformation ratio $n = N_1/N_2$. From transformer theory we have that $V_1/V_2 = I_2/I_1 = n$, so

$$V_1 = nV_2 \quad \text{and} \quad I_1 = I_2/n$$

These relationships may be written

$$V_1 = nV_2 + 0I_2 \tag{9.59}$$

$$I_1 = 0V_2 + (1/n)I_2 \tag{9.60}$$

Comparing Equations (9.59) and (9.33) we see that $A = n$ and $B = 0$. Comparing Equations (9.60) and (9.34) we see that $C = 0$ and $D = 1/n$. In matrix form

$$\begin{bmatrix} V_1 \\ I_1 \end{bmatrix} = \begin{bmatrix} n & 0 \\ 0 & 1/n \end{bmatrix} \begin{bmatrix} V_2 \\ I_2 \end{bmatrix} \tag{9.61}$$

Example 9.12

Obtain the *ABCD*-parameters for a practical transformer.

Solution

A practical transformer has resistance and leakage inductance associated with its windings and these are taken into account by lumped series impedances in series with each 'perfect' winding. For the purposes of analysis it is usual to refer the whole of this impedance to one side of the transformer.

A real transformer also has losses due to hysteresis and eddy currents and these are taken into account using a shunt conductance. This, together with a shunt susceptance used to take account of the need for a magnetizing current, gives a shunt admittance Y. The equivalent circuit then takes the form shown in Fig. 9.17.

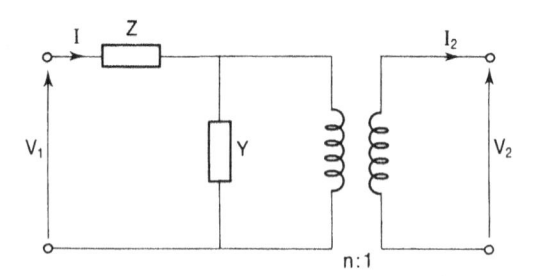

Figure 9.17

The diagram shows that the practical transformer equivalent circuit may be considered to be made up of a series impedance cascaded with a shunt admittance and then a perfect transformer. Using the transfer matrices developed in Examples 9.4, 9.7, and 9.11 we have

$$\begin{bmatrix} V_1 \\ I_1 \end{bmatrix} = \begin{bmatrix} 1 & Z \\ 0 & 1 \end{bmatrix} \begin{bmatrix} 1 & 0 \\ Y & 1 \end{bmatrix} \begin{bmatrix} n & 0 \\ 0 & 1/n \end{bmatrix} \begin{bmatrix} V_2 \\ I_2 \end{bmatrix}$$

Multiplying the first two transfer matrices we obtain

$$\begin{bmatrix} V_1 \\ I_1 \end{bmatrix} = \begin{bmatrix} 1 + ZY & Z \\ Y & 1 \end{bmatrix} \begin{bmatrix} n & 0 \\ 0 & 1/n \end{bmatrix} \begin{bmatrix} V_2 \\ I_2 \end{bmatrix}$$

Finally, multiplying the remaining two transfer matrices we get

$$\begin{bmatrix} V_1 \\ I_1 \end{bmatrix} = \begin{bmatrix} (1 + ZY)n & Z/n \\ Yn & 1/n \end{bmatrix} \begin{bmatrix} V_2 \\ I_2 \end{bmatrix} \tag{9.62}$$

9.9 CHARACTERISTIC IMPEDANCE (Z_o)

The characteristic impedance of a symmetrical two-port network is the impedance which, when connected to the output port, gives rise to an input impedance of the same value. This is illustrated in Fig. 9.18.

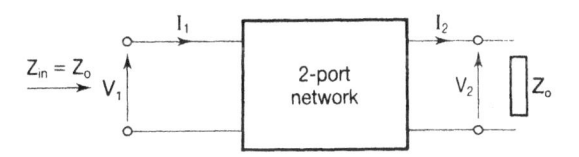

Figure 9.18

Using the transmission parameters,

$$V_1 = AV_2 + BI_2 = AV_2 + BV_2/Z_o$$

and

$$I_1 = CV_2 + BI_2 = CV_2 + DV_2/Z_o$$

The input impedance is

$$Z_{in} = V_1/I_1 = (AV_2 + BV_2/Z_o)/(CV_2 + DV_2/Z_o)$$

Multiplying throughout the numerator and the denominator of the right-hand side by Z_o/V_2 we obtain

$$V_1/I_1 = (AZ_o + B)/(CZ_o + D)$$

By definition, this is equal to Z_o

$$Z_o = (AZ_o + B)/(CZ_o + A)$$

remembering that $D = A$ for this symmetrical network. Thus

$$CZ_o^2 + AZ_o = AZ_o + B$$

so

$$CZ_o^2 = B$$

and

$$Z_o = \sqrt{(B/C)} \tag{9.63}$$

From Equations (9.33) and (9.34) we have, with the output port open circuited so that $I_2 = 0$,

$$V_1 = AV_2 \quad \text{and} \quad I_1 = CV_2$$

The open circuit input impedance $Z_{oc} = (V_1/I_1)|_{I_2=0} = (AV_2)/(CV_2)$, so

$$Z_{oc} = A/C \tag{9.64}$$

With the output port short circuited so that $V_2 = 0$,

$V_1 = BI_2$ and $I_1 = DI_2$

The short circuit input impedance is $Z_{sc} = (V_1/I_1)|_{V_2=0} = (BI_2)/(DI_2)$, so

$$Z_{sc} = B/D \qquad (9.65)$$

From Equations (9.64) and (9.65) we have that $Z_{oc}Z_{sc} = AB/CD$ and, for a symmetrical network, $A = D$, so

$Z_{oc}Z_{sc} = B/C$

Now from Equation (9.63), $Z_o = \sqrt{(B/C)}$, so

$$Z_o = \sqrt{(Z_{oc}Z_{sc})} \qquad (9.66)$$

Example 9.13

Determine (1) the *ABCD*-parameters and (2) the characteristic impedance of the network shown in Fig. 9.19.

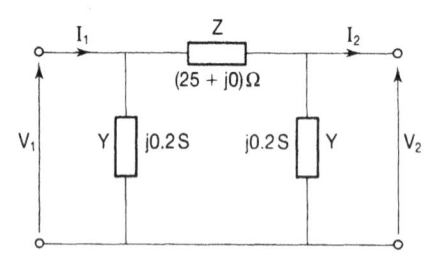

Figure 9.19

Solution

From matrix Equation (9.57) we have for the π-network:

$A = D = 1 + ZY = 1 + (25 \times j0.2) = 1 + j5$
$= \sqrt{(1^2 + 5^2)} \angle \tan^{-1}(5/1) = 5.09\angle 78.69°$
$B = Z = 25\ \Omega$
$C = 2Y + ZY^2 = j0.4 + (25)(j0.2)^2 = (-1 + j0.4)$ S

This means that C is in the third quadrant so that

$C = \sqrt{(1^2 + 0.4)^2}\angle[180 - \tan^{-1}(0.4/1)] = 1.07\angle 158.22°$ S

From Equation (9.63)

$Z_o = \sqrt{(B/C)} = \sqrt{[25/(1.07\angle 158.22°)]} = 4.83\angle -158.22°\ \Omega$

9.10 IMAGE IMPEDANCES

Suppose that for a *non-symmetrical* two-port network, terminating the output port with an impedance Z_{12} results in an impedance looking into the input terminals of Z_{11}, and terminating the input port with Z_{11} results in an impedance looking into the output terminals of Z_{12}. Then Z_{11} and Z_{12} are said to be image impedances. Networks are often designed on an image impedance basis in order to take advantage of the maximum power transfer theorem. For a symmetrical two-port network $Z_{11} = Z_{12}$ and is the characteristic impedance of the network, Z_o.

9.11 INSERTION LOSS

As we saw in Chapter 3, the maximum power transfer theorem tells us that maximum power is transferred from source to load when the impedance of the load is equal to the impedance of the source. When a network (an attenuator pad, for example) is inserted between a source and a load, as shown in Fig. 9.20, there will be a loss of power transfer due to the resulting mismatch, as well as that due to the loss in the inserted network itself.

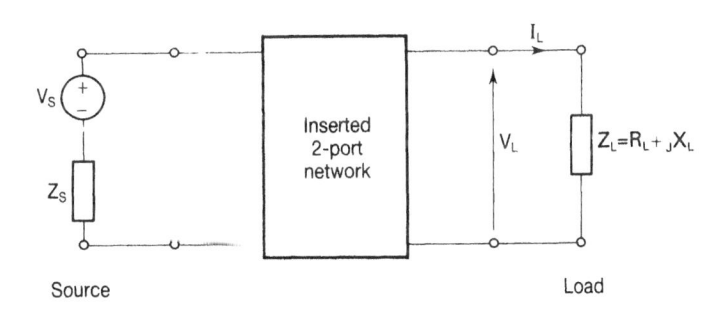

Figure 9.20

The insertion loss is defined to be

$$10 \log (P_b/P_a) \text{ dB} \tag{9.67}$$

where P_b is the power in the load before the network is inserted and P_a is the power in the load after the network is inserted. If the load resistance is R_L, the power in the load is $P = I^2 R_L$, so

$$P_b/P_a = I_b^2 R_L / I_a^2 R_L = I_b^2 / I_a^2$$

where I_b is the load current before the network is inserted and I_a is the load current after the network is inserted. It follows that the insertion loss is $10 \log (I_b^2/I_a^2)$, so

$$\text{insertion loss} = 20 \log (I_b/I_a) \text{ dB} \tag{9.68}$$

Example 9.14

The network shown in Fig. 9.21 is inserted between the generator and the load resistor shown in Fig. 9.22. Determine the insertion loss.

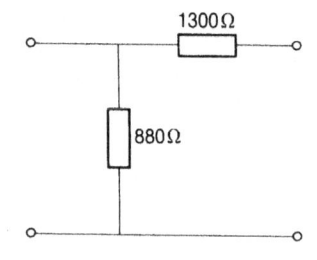

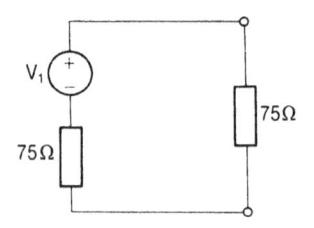

Figure 9.21

Figure 9.22

Solution

The network is now as shown in Fig. 9.23.

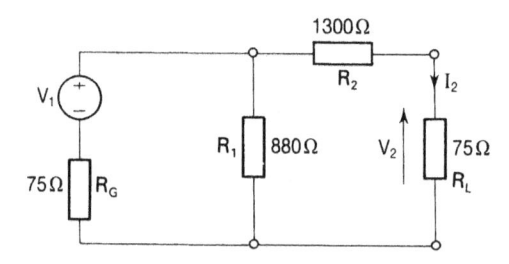

Figure 9.23

Before insertion we see from Fig. 9.22 that

$V_2 (=V_b \text{ say}) = [75/(75 + 75)]V_1 = 0.5V_1$

After insertion we see from Fig. 9.23 that $V_2 (=V_a \text{ say}) = I_2 R_L$. Now

$I_2 = [R_1/(R_2 + R_L + R_1)]I_1$

But

$I_1 = V_1/[R_G + \{R_1(R_2 + R_L)/(R_1 + R_2 + R_L)\}]$
$= V_1(R_1 + R_2 + R_L)/[R_G(R_1 + R_2 + R_L) + R_1(R_2 + R_L)]$

so

$I_2 = [R_1/(R_2 + R_L + R_1)][V_1(R_1 + R_2 + R_L)/\{R_G(R_1 + R_2 + R_L) + R_1(R_2 + R_L)\}]$
$= R_1 V_1/[R_G(R_1 + R_2 + R_L) + R_1(R_2 + R_L)]$

and

$V_a = I_2 R_L = R_L R_1 V_1/[R_G(R_1 + R_2 + R_L) + R_1(R_2 + R_L)]$

Putting in the values we have

$V_a = 75 \times 880 V_1/[75(880 + 1300 + 75) + 880(1300 + 75)]$
$= 0.0479 V_1$

From Equation (9.68), the insertion loss is $20 \log (I_b/I_a)$. This is equivalent to

$20 \log (I_b R_L / I_a R_L) = 20 \log (V_b / V_a)$

Thus the insertion loss is

$20 \log (0.5 V_1 / 0.0479 V_1) = 20.37$ dB

9.12 PROPAGATION COEFFICIENT (γ)

This is defined as the natural logarithm of the ratio of input to output currents or voltages when the network is terminated in its characteristic impedance. Thus

$$\gamma = \ln (I_1/I_2) \tag{9.69}$$

Also

$$\gamma = \ln (V_1/V_2) \tag{9.70}$$

As well as there being a change in level between I_1 and I_2 (or between V_1 and V_2) there will in general be a change in phase between them, so that γ will be complex. In general, therefore,

$$\gamma = \alpha + j\beta \tag{9.71}$$

From Equations (9.69) and (9.70) we see that

$I_1/I_2 = e^\lambda$

and

$V_1/V_2 = e^\lambda$

It follows that

$$I_1/I_2 = e^{\alpha+j\beta} = e^\alpha e^{j\beta} = e^\alpha \angle \beta \tag{9.72}$$

Similarly

$$V_1/V_2 = e^\alpha \angle \beta \tag{9.73}$$

where

$$\alpha = \ln |I_1/I_2| \text{ nepers } (= \ln |V_1/V_2| \text{nepers}) \tag{9.74}$$

and is called the attenuation coefficient because it is responsible for the change in level between input and output quantities.

β (measured in radians or degrees) is called the phase change coefficient because it gives the change in phase between the input and output quantities.

9.13 SELF-ASSESSMENT TEST

1. Define a two-port network.
2. Give two examples of a two-port network.
3. Give another name for a two-port network.
4. Why are the impedance parameters of a two-port network called the open circuit parameters?
5. Which parameters of a two-port network are also known as the short circuit parameters?
6. For what purpose are the h-parameters commonly used?
7. What are the inverse h-parameters also known as?
8. Give another name for the $ABCD$-parameters of a two-port network.
9. Explain what is meant by a symmetrical two-port network.
10. Which two of the $ABCD$-parameters are always equal in symmetrical two-port networks?
11. Give an example of a symmetrical two-port network.
12. Give an example of an unsymmetrical two-port network.
13. Draw a diagram to represent two two-port networks connected in cascade.
14. Explain what is meant by the characteristic impedance (Z_o) of a two-port network.
15. Give an expression for the characteristic impedance of a two-port network in terms of one or more of its $ABCD$-parameters.
16. Give an expression for the characteristic impedance of a two-port network in terms of its open and short circuit impedances.
17. Explain what is meant by 'image impedances'.
18. Explain what is meant by the term 'insertion loss' as applied to two-port networks.
19. Give an expression for the insertion loss of a two-port network in terms of load currents.
20. Define the term 'propagation coefficient' (γ) as applied to two-port networks.
21. In general the propagation coefficient is complex and is given by $\gamma = \alpha + j\beta$. What is the significance of α and β?

9.14 PROBLEMS

1. Viewed from the input port a four-terminal network consists of a series impedance of 300 Ω followed by a shunt admittance of (1/900) S. Determine (a) the z-parameters and (b) the y-parameters of the network.

2. A two-port T-network has a series resistance of 300 Ω followed by a shunt conductance of (1/900) S and then a series resistance of 600 Ω. Determine (a) the z-parameters and (b) the y-parameters of the network.

3. A transistor has the following h-parameters: $h_{11} = 1$ kΩ; $h_{12} = 10^{-3}$; $h_{21} = 100$; $h_{22} = 10^{-4}$ S. Using the equivalent circuit of Fig. 9.5 determine (a) the voltage gain (V_2/V_1) and (b) the current gain (I_2/I_1) when a load resistance of 1 kΩ is connected across the output terminals.

4. Determine the g-parameters of the transistor of Problem 3.

5. Each phase of a transmission line has a total series impedance of $95\angle 75°$ Ω and a shunt admittance of $1.04\angle 90°$ mS. Obtain the ABCD-parameters for the nominal-π representation of the line.

6. A two-port T-network has series impedances $Z_1 = 2\angle 60°$ Ω and $Z_2 = 5\angle 70°$ Ω and a shunt admittance of $0.01\angle 80°$ S. Obtain the ABCD-parameters for the line.

7. Five two-port networks consisting of a series resistance of 100 Ω, a shunt conductance of (1/400) S, a series resistance of 200 Ω, a shunt conductance of (1/400) S and a series resistance of 100 Ω are connected in cascade in that order. Obtain the ABCD-parameters of the cascaded network.

8. A symmetrical-T four-terminal network has series resistances R_1 and R_2 each of 400 Ω and a shunt branch of resistance 600 Ω. Determine the characteristic impedance of the network.

9. Show that for non-symmetrical two-port networks the image impedances are given by $\sqrt{[(AB)/(CD)]}$ and $\sqrt{[(DB/CA)]}$ where A, B, C and D are the transmission parameters. [Hint: for each expression use the same method as was used to obtain the expression $Z_o = \sqrt{(B/C)}$, Equation (9.63).]

10. If the resistor R_2 in Problem 8 is replaced with one of 600 Ω resistance, determine (a) the ABCD-parameters of the new network and (b) the image impedances Z_{11} and Z_{12}.

11. An attenuator pad has series resistances each of 200 Ω and a shunt resistance of 800 Ω in a T-section. Calculate the insertion loss when it is inserted between a load resistor of 1 kΩ and a 30 V source having an internal resistance of 1 kΩ.

12. The network of Problem 8 is inserted between a source and a load. Determine the insertion loss if the resistance of the source and the load

13 Each series arm of a symmetrical low-pass filter consists of a pure inductor of 0.018 H and the shunt branch is a capacitor of 100 μF capacitance. Determine the characteristic impedance of the network when operating at (a) 1 kHz and (b) 8 kHz.

10 Duals and analogues

10.1 DUALS OF CIRCUIT ELEMENTS

We have seen that in linear circuit theory there is an intimate relationship between voltage and current. Their relationship is expressed in terms of impedance or admittance by the following equations:

$$V = IZ \tag{10.1}$$

$$I = VY \tag{10.2}$$

These equations ultimately give the same information and the operations involved in solving for V or I are the same. Each equation is said to be the dual of the other. The elements of the equations are similarly dual pairs so that voltage is the dual of current and impedance is the dual of admittance. The component parts of impedance are resistance and inductive (or capacitive) reactance whose duals are, respectively, conductance and capacitive (or inductive) susceptance. The dual of $(R + jX_L)$ is thus $(G - jB_C)$ and the dual of $(R - jX_C)$ is $(G + jB_L)$.

Given an equation, therefore, its dual can immediately be written down by replacing each one of its component parts by its dual. Table 10.1 shows the duals of the circuit elements.

Table 10.1

Quantity	←→	Dual
Voltage		Current
Impedance		Admittance
Resistance		Conductance
Inductance		Capacitance

Example 10.1

Obtain the dual of the expression for the energy stored in a capacitor of capacitance C across which is maintained a voltage V.

Solution

The energy stored in the capacitor is given by $W = (CV^2)/2$. The dual of capacitance is inductance and the dual of voltage is current. Replacing the component parts of the above equation by their duals we get, for the required dual expression, that the energy W stored in an inductor of inductance L through which is flowing a current I is given by $W = (LI^2)/2$.

10.2 DUAL CIRCUITS

The circuits described by Equations (10.1) and (10.2) are shown in Fig. 10.1(a) and (b), respectively. In Fig. 10.1(a) the voltage is the source or stimulus, and the current through the impedance is the circuit response, whereas in the circuit of Fig. 10.1(b) the current is the source or stimulus, and the voltage across the admittance is the circuit response. Notice that the two parts of the previous sentence are dual statements.

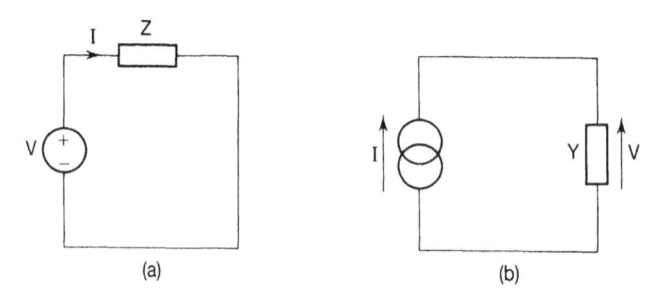

Figure 10.1

Series and parallel circuits

If we have a series circuit such as that shown in Fig. 10.2(a) for which $V = I(Z_1 + Z_2 + Z_3 + Z_4)$, the dual equation is obtained by replacing the elements of the equation by their duals so that we get

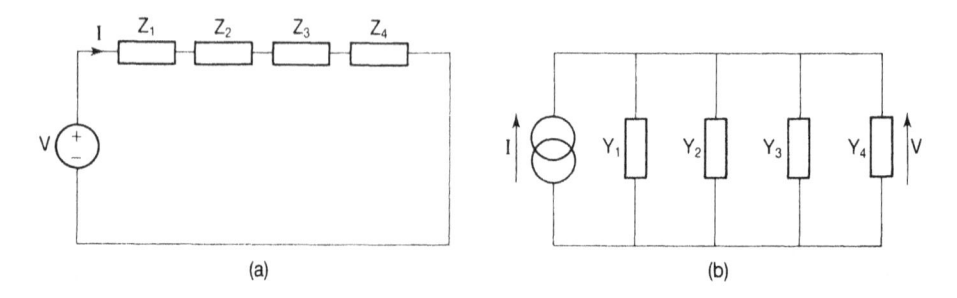

Figure 10.2

$I = V(Y_1 + Y_2 + Y_3 + Y_4)$ and the circuit described by this equation is shown in Fig. 10.2(b), which is thus the dual of that in Fig. 10.2(a).

It follows that:

- a parallel circuit is the dual of a series circuit;
- 'impedances in series are added' and 'admittances in parallel are added' are dual statements; and
- 'when elements are in series, voltages are added' and 'when elements are in parallel currents are added' are dual statements.

Kirchhoff's current law and Kirchhoff's voltage law

From the circuits of Fig. 10.2 we see that the voltage V is the sum of the voltages across the impedances Z_1, Z_2, Z_3 and Z_4 (which is KVL), while the current I is the sum of the currents in Y_1, Y_2, Y_3 and Y_4 (KCL). Thus Kirchhoff's current law is the dual of his voltage law.

Nodal voltage and mesh current

It follows from the previous paragraph that the dual of a closed path (a loop or mesh) is a node and that mesh current analysis and nodal voltage analysis are dual procedures.

Thevenin's theorem and Norton's theorem

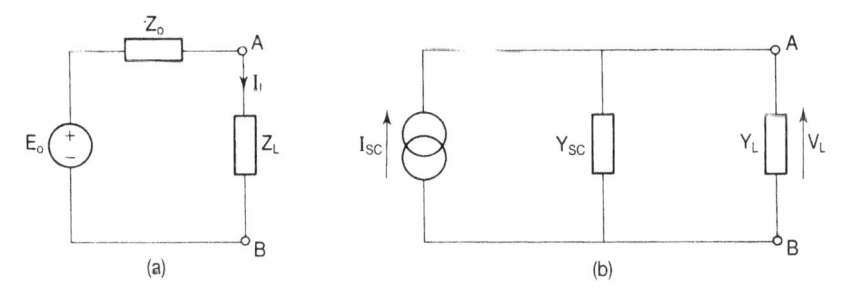

Figure 10.3

The Thevenin equivalent circuit of Fig. 10.3(a) consists of an open circuit voltage E_o in series with an impedance Z_o. The current through the load impedance Z_L connected across the output terminals A and B is then calculated from the equation

$$I_L = E_o/(Z_o + Z_L) \tag{10.3}$$

The dual of a voltage source is a current source and the dual of a series impedance is a parallel admittance. The dual of the circuit of Fig. 10.3(a) is thus that of Fig. 10.3(b), which is the Norton equivalent circuit. This circuit consists

236 Duals and analogues

of a short circuit current in parallel with an admittance Y_{SC}. The voltage across the load admittance Y_L, connected across the load terminals A and B, is then calculated from the equation

$$V_L = I_{SC}/(Y_{SC} + Y_L) \tag{10.4}$$

Notice the duality of Equations (10.3) and (10.4) and of the statements in the above two paragraphs.

Open circuit and short circuit

It can be seen from the consideration of Thevenin's and Norton's theorems that a short circuit is the dual of an open circuit.

Example 10.2

Obtain the duals of the circuits shown in Fig. 10.4.

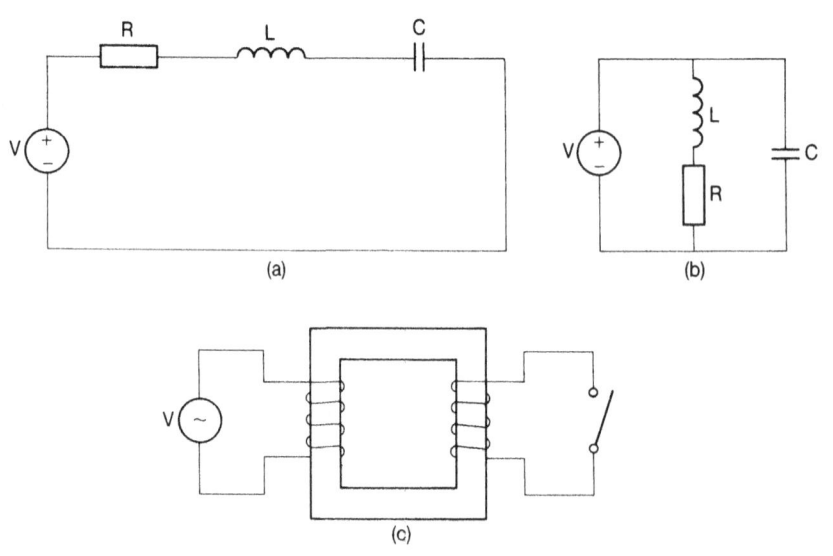

Figure 10.4

Solution

(a) The voltage source is replaced by a current source; the circuit consisting of a resistor R, an inductor L and a capacitor C in series is replaced by one consisting of a conductance G, a capacitor C and an inductor L in parallel. This is shown in Fig. 10.5(a).

(b) The voltage source is replaced by a current source; the circuit consisting of a series combination of a resistor R and an inductor L in parallel with a capacitor C is replaced with one consisting of a parallel combination of a conductance G and a capacitor C in series with an inductor L. This is shown in Fig. 10.5(b).

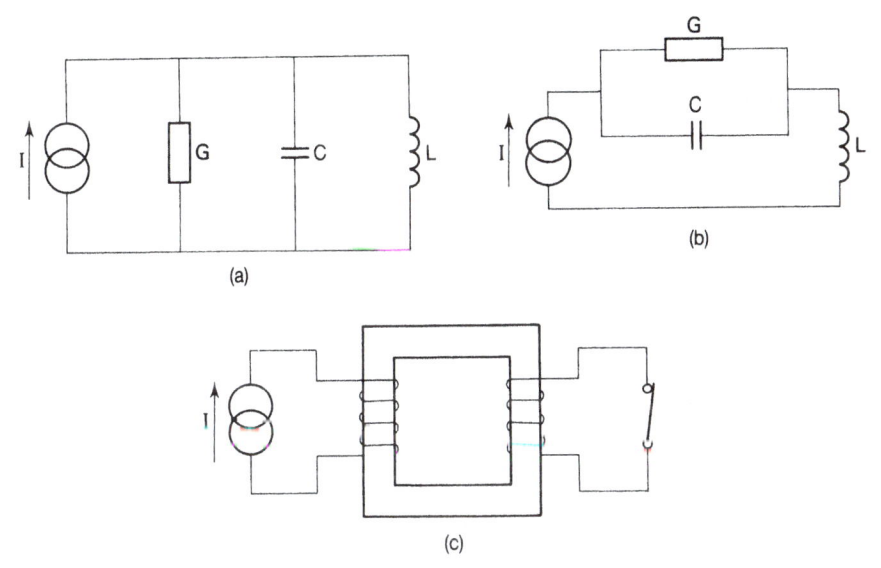

Figure 10.5

(c) This circuit shows an ideal transformer whose secondary winding is open circuited and whose primary winding is fed from a voltage source. The dual circuit will have the primary winding fed from a current source and its secondary winding will be on short circuit. This is shown in Fig. 10.5(c). Notice that, whereas the ultimate disaster in the circuit of Fig. 10.4(c) would be a short circuited secondary winding leading to infinite current, the corresponding catastrophe for the transformer in the dual circuit of Fig. 10.5(c) would be an open circuited secondary winding leading to infinite voltage.

A summary of the dual pairs associated with electric circuits is given in Table 10.2.

It must be emphasized that dual circuits are not equivalent circuits. Their usefulness lies in modelling of systems. For example, it is much easier to obtain a capacitor with a leakage resistance tending to infinity than it is to obtain an inductor with a resistance tending to zero. In modelling systems having inductors, therefore, dual circuits can be used to represent the 'real' system.

Table 10.2

Circuit statement	Dual
Series	Parallel
Series impedances are added	Parallel admittances are added
Open circuit	Short circuit
Switch open	Switch closed
Node	Loop/mesh
KCL	KVL
Thevenin's theorem	Norton's theorem
Nodal voltage analysis	Mesh current analysis

10.3 ANALOGUES

If an equation describing the operation of a physical system is identical in form to one describing the operation of another physical system, then the corresponding quantities in each equation are said to be analogous. The equations themselves are also analogous. The form of the equations and the associated mathematical manipulation are the important consideration, not the physical similarity (or otherwise) between the systems.

There are two major advantages of analogous systems. One is the saving in memory space resulting from the form of equations being identical in two or more systems. The second is that, when dealing with mechanical/electrical analogues, for example, it is possible to express the whole system in an integrated form mathematically.

Electric, magnetic and conduction fields

The field vectors of the electric field are related by the equation

$$D = \epsilon E \qquad (10.5)$$

where D is the electric flux density, E is the electric field strength and ϵ is the permittivity of the medium of the field.

Similarly, for the magnetic field we have

$$B = \mu H \qquad (10.6)$$

where B is the magnetic flux density, H is the magnetic field strength and μ is the permeability of the medium of the field.

For the conduction field

$$J = \sigma E \qquad (10.7)$$

where J is the current density, E is the electric field strength and σ is the conductivity of the medium of the field.

We notice that Equations (10.5), (10.6) and (10.7) are identical in form and are said to be analogous. Similarly the corresponding quantities (D, B and J; E and H; ϵ, μ and σ) in each of the equations are analogues. Any one of these equations could be obtained from one of the others by replacing each quantity of the second system by the corresponding analogue from the first.

Example 10.3

The energy stored in an electric field is given, in terms of the field vectors, by the equation $W = DE/2$ joules per cubic metre of the field. By consideration of field analogies obtain an expression for the energy stored per cubic metre in a magnetic field.

Solution

The magnetic field analogues of D and E are, respectively, B and H so that the analogous energy equation is $W = BH/2$ joules per cubic metre.

Electric and magnetic circuits

It is often convenient to use the analogies between electric and magnetic circuits when analysing the latter. For example, the series–parallel magnetic circuit shown in Fig. 10.6(a) is the analogue of the series–parallel electric circuit shown in Fig. 10.6(b).

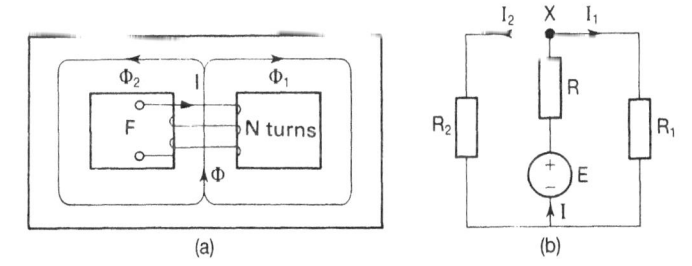

Figure 10.6

The resistance of an electrical conductor (copper, for example) is given by $R = l/\sigma A$ where l is the length of the conductor, A is its cross-sectional area, and σ is the conductivity of the copper. The reluctance of a magnetic 'conductor' (iron, for example) is given by $S = l/\mu A$ where l is the length of the iron path, A is its cross-sectional area and μ is the permeability of the iron. Resistance is a measure of how difficult it is for current to flow in an electric circuit and reluctance is a measure of how difficult it is for magnetic flux to 'flow' in a magnetic circuit. Resistance and reluctance are analogues of one another.

In the circuits of Fig. 10.6, the resistance R_1 of the electric circuit is analogous to the reluctance S_1 of the right-hand limb of the magnetic circuit. Similarly, resistors R_2 and R are analogous to S_2 and S (respectively, the reluctances of the left-hand limb and the centre limb of the magnetic circuit). The electromotive force (emf) E in the central branch of the electric circuit is analogous to the magnetomotive force (mmf) $F (=NI)$ in the centre limb of the magnetic circuit. Finally, the currents I_1, I_2 and I are the electric circuit analogues of the fluxes Φ_1, Φ_2 and Φ, respectively, in the magnetic circuit.

Applying KVL to the left-hand mesh gives

$$E = IR + I_2 R_2 \tag{10.8}$$

Applying KVL to the right-hand mesh gives

$$E = IR + I_1 R_1 \tag{10.9}$$

Applying KVL to the outer loop gives

$$I_1 R_1 = I_2 R_2 \tag{10.10}$$

Equation (10.10) could of course been obtained from Equations (10.8) and (10.9) simply by equating their right-hand sides.

Applying KCL to node X gives

$$I = I_1 + I_2 \tag{10.11}$$

By analogy, the equations for the magnetic circuit may now be written down immediately. Thus, from Equation (10.8) we have

$$F(=NI) = \Phi S + \Phi_2 S_2 \tag{10.12}$$

From Equation (10.9) we see that

$$NI = \Phi S + \Phi_1 S_1 \tag{10.13}$$

Equation (10.10) indicates that

$$\Phi_1 S_1 = \Phi_2 S_2 \tag{10.14}$$

Finally, by analogy with Equation (10.11), we have

$$\Phi = \Phi_1 + \Phi_2 \tag{10.15}$$

Table 10.3 summarizes the analogies between the electric and magnetic fields.

Table 10.3

Electric		Magnetic		Conductive	
Electric flux,	ψ	Magnetic flux,	Φ	Electric current,	I
Field strength,	E	Field strength,	H	Field strength,	E
Flux density,	D	Flux density,	B	Current density,	J
Permittivity,	ϵ	Permeability,	μ	Conductivity,	σ
		Reluctance,	S	Resistance,	R
		mmf,	F	emf,	E

Electrical and mechanical systems

Electric circuits are made up of energy sources, sinks and stores represented, respectively, by voltage or current sources, resistors and inductors or capacitors. Similarly, in mechanical systems there are sources of force, together with energy sinks (for example, dashpots) and energy stores (for example, springs or moving masses). There are analogous quantities in the two systems leading to analogous equations. There are summarized in Table 10.4.

Since the form of the equations shown in the table is identical in both systems, their manipulation is likewise identical.

Table 10.4

Mechanical	Electrical analogue
Force source, P	Current source, I
Velocity, v	Voltage, V
Viscous resistance, B	Conductance, G
Spring compliance, D	Inductance, L
Mass, M	Capacitance, C
$P = Bv$	$I = GV$
$v = DdP/dt$	$V = LdI/dt$
$P = Mdv/dt$	$I = CdV/dt$

10.4 SELF-ASSESSMENT TEST

1 State the dual of voltage.

2 Give the dual equation of $V = IZ$.

3 Give the dual expression of $(CV^2)/2$.

4 Give the dual of a circuit containing a resistor and a capacitor connected in parallel.

5 State the dual of Norton's theorem.

6 Give the magnetic field analogue of permittivity.

7 State the conduction field analogue of magnetic flux density.

8 What is the magnetic field analogue of electric current?

9 Give the magnetic circuit equation which is analogous to $E = IR$.

10 The expression for the capacitance C of a parallel plate capacitor having plate area A and separation d is $C = A\epsilon/d$, where ϵ is the permittivity of the dielectric material. Use the method of field analogues to write down the expression for the conductance of the dielectric material.

11 'Current cannot change instantaneously in an inductor'. What is the dual statement in relation to capacitors?

12 Give the mechanical system analogue of electrical inductance.

13 Give the electrical system analogue of the mechanical system equation $P = Bv$ where P is the force source, B is the viscous resistance, and v is velocity.

14 When a step voltage is applied to an RL circuit the current as a function of time is given by $i = I[1 - \exp(-Rt/L)]$. Give the dual equation in respect of a step voltage being applied to an RC circuit.

Answers to self-assessment tests and problems

Chapter 1

Self-assessment test

1. second; newton; kilogram. **2.** coulomb; newton; volt. **3.** $[M\,L^2\,T^{-2}\,A^{-1}]$. **4.** $[M\,T^{-2}\,A^{-1}]$.
5. $[M\,L\,T^{-3}\,A^{-1}]$. **6.** (a) 30×10^{-3} A; (b) 25×10^6 µA; (c) 10×10^9 mW; (d) 25×10^{-9} C; (e) 150×10^{-3} nF; (f) 60×10^{-3} GW; (g) 150×10^{-3} mJ; (h) 220×10^{-3} kΩ; (i) 55×10^9 mΩ; (j) 100×10^{-3} kN.

Problems

1. $[L^{-1}\,A]$. **2.** $[M\,L\,T^{-2}\,A^{-2}]$. **3.** $[T\,A]$. **4.** $[L^{-2}\,T\,A]$. **5.** $[M\,L\,T^{-3}\,A^{-1}]$. **6.** True.
7. $[M^{-1}\,L^{-3}\,T^4\,A^2]$. **8.** $a = 2$; $b = 1$. **9.** False, $W = (L\,I^2)/2$. **10.** $a = 2$; $b = -1$; $c = -1$.
11. $a = 2$; $b = -1$; $c = -2$. **12.** 0.145 MW.

Chapter 2

Self-assessment test

8. (1) 3.33 Ω, (2) 30 Ω, (3) 15 Ω. **9.** 0.15 Ω. **11.** Between 198 Ω and 242 Ω. **19.** 16 µH.
20. 82 µH.

Problems

1. 0.5 Ω. **2.** 64 Ω. **3.** 4.04 Ω. **4.** 20 V. **5.** 5 A. **6.** 1 Ω. **7.** 59.7 °C. **8.** red, red, red, silver.
9. Between 278 Ω and 338 Ω. **10.** 4.29 µF, 7.14 µF, 8.57 µF. **11.** 125 µH, 80 µH, 100 µH.
12. 405 µH, 5 µH. **13.** 0.919. **14.** 9 A. **15.** 2 mJ. **16.** 1.44 kW.

Chapter 3

Self-assessment test

9. 4 Ω. **10.** 30 Ω. **11.** 100 Ω, 900 Ω. **12.** 110 Ω, 0 Ω, 0 Ω.

Problems

1. 1.49 A. **2.** 3.64 A. **3.** $E_o = 30$ V, $R_o = 2$ Ω. **4.** 720 mA (maximum), 11.5 mA (minimum).
5. (a) 440 mA, (b) 9.87 mA, (c) 17.5 Ω. **6.** $I_1 = 224$ mA, $I_2 = 188$ mA, $I_3 = 8$ mA,

$I_4 = 44$ mA. **7.** (a) 60 Ω, (b) 8 A, 4 A, 4 A. **8.** 0.332 Ω, 66.4 Ω, 0.664 Ω. **9.** (a) 2 Ω, (b) 112.5 W. **10.** (a) 1.4 V, 3.9 Ω; (b) 1.28 V. **11.** (a) 0.456 A flowing from A to B, (b) 11.5 Ω in parallel with R, (c) 2.29 W. **12.** 3.79 mA.

Chapter 4

Self-assessment test

2. Hz. **3.** $f = 1/T$. **4.** 25. **5.** 50 Hz. **6.** 90°. **7.** 100 V. **8.** 1.11. **9.** In phase. **10.** 150.8 Ω. **11.** 31.83 kΩ. **12.** Ohm. **13.** 63.43°. **14.** 28.28 Ω. **15.** (a) 22.36 Ω, (b) 20 Ω. **17.** j. **18.** Third. **19.** 18∠−56.3°. **20.** 21.65 + j12.5. **21.** 5∠53.13°. **22.** (5 + j6) Ω. **23.** (0.1 + j0.04) S. **24.** $S = P + jQ$. **25.** (a) 160 VA, (b) 125 W, (c) 100 Var, (d) 0.832 leading, (e) (25 − j20) V.

Problems

1. $v = 250 \sin (314t + 53.2°)$; $i = 5 \sin (314t + 23.6°)$; −150 V, −4.58 A; 74.6° current lagging. **2.** 122 V leading v_1 by 2.2°. **3.** (a) 5 Ω, 5 Ω, 7.07 Ω; 5 Ω, 10 Ω, 11.2 Ω. **4.** (a) 11.1 A, (b) 0.555 lagging. **5.** (a) 2.47 A, (b) 123.5 V, (c) 157.3 V. **6.** (a) 7.98 Hz, (b) 433 Ω. **7.** (a) (i) 2.48 A, (ii) 124 V, 158 V, (iii) 0.618 leading; (b) 36.7 μF; (c) 110 μF. **8.** (a) 2 A, 1.57 A; (b) 2.54 A; (c) 78.8 Ω; (d) 38° leading. **9.** (a) (i) 13.2 A, (ii) 2.62 kW, (iii) 0.991 lagging; (b) (i) 31.5 A, (ii) 5.9 kW, (iii) 0.937 lagging. **10.** (a) 0.118 S, 0.113 S, 0.164 S; (b) 32.8 A; (c) 4.72 kW. **11.** (a) 13.1 A, (b) 0.953 leading. **12.** (a) 23.6 A, (b) 5.1 kW.

Chapter 5

Self-assessment test

7. ABC. **9.** $E_M \sin (\omega t - 240°)$. **11.** 25 A. **12.** 30°. **13.** 17.32 A. **14.** 12.65 kW. **15.** $\sqrt{3} V_L I_L$. **16.** 28 kVar. **17.** 340 W. **18.** $W_1 = V_{AC} I_A \cos$ (angle between V_{AC} and I_A); $W_2 = V_{BC} I_B \cos$ (angle between V_{BC} and I_B). **19.** $\cos \{\tan^{-1}[\sqrt{3}(P_1 - P_2)/(P_1 + P_2)]\}$. **20.** (a) 190 W, (b) 103.92 W, (c) 216.56 W, (d) 0.877.

Problems

1. (a) 32.44 A, (b) 8.99 kW. **2.** (a) 34.6 A, (b) 18.24 kW. **3.** (2.287 − j10.79) kVA. **4.** (a) 30 kW, (b) 0.33, (c) 231 A. **5.** 0.866. **6.** (a) 400 kW, (b) 0.756, (c) 152 A, (d) 360 kW.

Chapter 6

Self-assessment test

1. $\omega L = 1/\omega C$. **2.** $1/2\pi\sqrt{(LC)}$. **3.** Unity. **4.** 250 rad s^{-1}. **5.** 2 Ω. **6.** Because the current is a maximum at resonance. **8.** Q is dimensionless. **9.** By reducing the resistance in the circuit. **12.** Hz. **14.** P_1 is $10 \log_{10}(P_1/P_2)$ dB above P_2. **16.** $1/2\pi\sqrt{(LC)}$. **17.** L/CR. **18.** It increases. **19.** A minimum. **20.** ω_0/Q.

Problems

1. 4 Ω, 504 mH. **2.** (a) 40 Hz, (b) 20 A, (c) 5 kV. **3.** (a) 20.3 μF, (b) 3.925 kV, 100 V, (c) 39.3. **4.** (a) 22.5 kHz, (b) 17.6, (c) 1.28 kHz. **5.** 0.233; −12.6 dB. **6.** (a) $1/[1 + 1 j\omega CR]$, (b) −40, −20, −3; Comment: 1000 rad s^{-1} is a half-power frequency. (c) Add a series inductance, the output still being taken across the resistor. **7.** (a) 38 μF, (b) 60.8 W, (c) 3.93 (11.89 dB).

8. (a) 38.5 Hz, (b) 50.23 Hz, (c) 1 A, 0.1 A. **9.** (a) 15.9 kHz, (b) 20, (c) 5 kHz, (d) 13.4 kHz, 18.4 kHz. **10.** (a) 33.7 µF, (b) 78 µF. **11.** (a) 12.8 kΩ, (b) 6.4 kΩ, (c) 0.127 µF, (d) 223, 2.8 kHz. **12.** (a) 50 kHz, (b) 7.95 kΩ, (c) 12.7 Ω. **13.** (a) 2.8 nF, (b) 7.14 kΩ, (c) 37.7, (d) 41.2 mA.

Chapter 7

Self-assessment test

3. [1 7 9 81]. **4.** a_{34}. **5.** Yes. **6.** No.

Problems

1. (a) $\begin{bmatrix} 2 & 3 \\ 4 & 4 \end{bmatrix}$, (c) $\begin{bmatrix} 0 & 3 \\ 6 & 8 \end{bmatrix}$, (c) $\begin{bmatrix} -2 & -6 \\ -1 & -12 \end{bmatrix}$, (d) $\begin{bmatrix} 1 & 3 \\ -11 & -15 \end{bmatrix}$

2. (a) $\begin{bmatrix} 2 & -3 \\ 1 & 5 \end{bmatrix} \begin{bmatrix} x \\ y \end{bmatrix} = \begin{bmatrix} 1 \\ 7 \end{bmatrix}$

(b) $\begin{bmatrix} 1 & 1 & 3 \\ 1 & -2 & 1 \\ 0 & -4 & -3 \end{bmatrix} \begin{bmatrix} x \\ y \\ z \end{bmatrix} = \begin{bmatrix} 7 \\ 6 \\ 4 \end{bmatrix}$

3. (a) $2x + 5y = 27$ (b) $x - y + 5z = 6$
$x + 3y = 16$ $x + y + z = 0$
                       $-x \quad\quad - 3z = 1$

4. 4. **5.** 25. **6.** 5. $\begin{vmatrix} 7 & 6 \\ 8 & 9 \end{vmatrix} = 15$. **7.** $(-1)^{2+1} \begin{vmatrix} 2 & 6 \\ 1 & 4 \end{vmatrix} = -2$. **8.** 3.64 A. **9.** 0.94 A. **10.** 1.2 A

11. −20.6 mA. **12.** (a) 0 A, 2 A, 4 A, (b) 52 W. **13.** 13.9 A

Chapter 8

Self-assessment test

6. τ, second. **7.** L/R. **8.** 10 ms. **9.** 10^3 A s^{-1}. **10.** 2 A. **11.** Exponential rise. **12.** Exponential decay. **13.** The resistor. **14.** The resistor. **15.** 500 µs. **16.** 5 ms. **17.** 0.362 A. **18.** 50 µs.

Problems

1. (a) 10 A s^{-1}, (b) 0.948 A, (c) 1 s, (d) 10 A, (e) 5 s. **2.** (a) 526.6 Ω, 2.11 H; (b) 94.97 As^{-1}. **3.** (a) 6.97 A; (b) 1 s. **4.** 2.72 MΩ. **5.** 2.89 MΩ. **6.** 17 ms. **7.** (b) $v_C = 10[1 - \exp(-1000t)]$, $i = 0.5[1 + \exp(-1000t)]$. **8.** (a) An exponential growth reaching 20 V after 5τ ($\tau = 2.55$ ms); (b) an exponential growth reaching 17.2 V after 5 ms and subsequently decaying exponentially to zero after a further 12.75 ms; (c) as for (b) but repeating 7.25 ms after becoming zero. **9.** (a) An exponential decay starting at 15 V and falling to zero in 15 µs; (b) an exponential decay starting at 15 V and falling to 2.8 V after 5 µs at which time the pulse is removed, then instantaneously reversing to −12.2 V and rising exponentially to reach zero after a further 15 µs; (c) as for (b) but repeating 5 µs after becoming zero; (d) an exponential decay starting at 15 V and falling to 2.8 V after 5 µs, then an instantaneous change to −12.2 V and an exponential rise towards zero reaching −2.3 V after a further 5 µs; another instantaneous change to +12.7 V, followed by an exponential decay to 2.39 V after a further 5 µs. This continues with each successive positive voltage becoming smaller

and each successive negative voltage becoming numerically larger.
10. $i(t) = (2/\omega) \exp(-\alpha t) \sin \omega t$ where $\alpha = r/2L = 15^5$ s and $\omega = \sqrt{[(1/CL) - (R/2L)^2]} = 1.73 \times 10^5$ rad s^{-1}. This is an exponentially decaying sine wave and is underdamped.

Chapter 9

Self-assessment test

3. A four-terminal network. **4.** Because they are measured with either the input or the output port on open circuit. **5.** The admittance (y) parameters. **6.** For the analysis of transistor circuits. **7.** g-parameters. **8.** Transmission parameters. **9.** The input and output ports may be interchanged without effect. **10.** A and D. **15.** $\sqrt{(B/C)}$. **16.** $\sqrt{(Z_{oc}Z_{sc})}$. **19.** $20 \log_{10}(I_b/I_a)$ dB where I_b is the load current before insertion and I_a is the load current after insertion.

Problems

1. (a) $z_{11} = 1200$ Ω, $z_{12} = z_{21} = z_{22} = 900$ Ω; (b) $y_{11} = (1/300)$ S, $y_{12} = y_{21} = (-1/300)$ S, $y_{22} = (4/900)$ S. **2.** (a) $z_{11} = 1200$ Ω, $z_{12} = z_{21} = 900$ Ω, $z_{22} = 1500$ Ω, (b) $y_{11} = (5/3300)$ S, $y_{12} = y_{21} = (-1/1100)$ S, $y_{22} = (4/3300)$ S. **3.** (a) -83, (b) -91. **4.** $g_{11} = 10^{-3}$ S, $g_{12} = 10^3$, $g_{21} = 10^{-2}$, $g_{22} = 10$ kΩ. **5.** $A = D = 0.95\angle 0.78°$, $B = 95\angle 75°$ Ω, $C = 1.01\angle 90.4°$ S. **6.** $A = 0.986\angle 0.78°$, $B = 7.06\angle 66.7°$ Ω, $C = 0.01\angle 80°$ S, $D = 0.957\angle 1.5°$. **7.** $A = D = 17/8$, $B = (4500/8)$ Ω, $C = (5/800)$ S. **8.** 775 Ω. **10.** (a) $A = 5/3$, $B = 1600$ Ω, $C = (1/600)$ S, $D = ?$; (b) $Z_{11} = 895$ Ω; $Z_{12} = 1072$ Ω. **11.** -6.4 dB. **12.** (a) 13 dB; (b) 9.5 dB. **13.** (a) 580 Ω; (b) j684 Ω.

Chapter 10

Self-assessment test

1. Current. **2.** $I = VY$. **3.** $(LI^2)/2$. **4.** A resistor and an inductor in series. **5.** Thevenin's Theorem. **6.** Permeability. **7.** Current density. **8.** Magnetic flux. **9.** $F(=NI) = S$. **10.** $G = A\sigma/d$. **11.** 'Voltage cannot change instantaneously in a capacitor'. **12.** Compliance. **13.** $I = GV$. **14.** $v_C = V[1 - \exp(-t/CR)]$.

Index

ABCD parameters 213
   of a π-network 220
   of a T-network 222
a.c. circuits
   purely capacitive 75
   purely inductive 73
   purely resistive 72
   parallel circuits 91
   series circuits 77
   series–parallel circuits 95
   single phase 72
   three phase circuits 107
admittance 92
   parameters 208
   triangle 92
alternating quantity 66
   average value 71
   instantaneous value 67
   maximum value 67
   peak value 67
   root mean square (rms) value 70
   sinusoidal a.c. 67
ampere 1
analogue 238
Argand diagram 83
attenuation coefficient 229

balanced 3-phase system 109
bandwidth 130, 135
branch 41
bridged-T circuit 58

capacitance 24
  of parallel plates 25
  of concentric cylinders 25
  of parallel cylinders 26
capacitors 25
  charging of 181
  discharging of 184
  energy stored in 35
  in parallel 27
  in series 26
characteristic impedance 225
charge 10
circuit element 11
  active 11
  passive 11
coefficient of coupling 31
complex notation 82
  application to a.c. circuit analysis 89
complex quantity
  addition and subtraction of 86
  multiplication and division of 88
conductance 15
conductivity 15
coulomb 10
coupled circuits 30
Cramer's Rule 146
current 12
  division 20
  source 12
cycle 66

damping
  critically damped 201
  overdamped 201
  underdamped 201
dB notation 131
decibel 131
delta connection 56, 114
delta star transformation 56
determinant 143
dielectric 25

differentiator
    RL   179
    RC   189
dimensional analysis   4
dimensions   1
discharging a capacitor   184
dot notation   31
dual circuit   234
duals   230
    of circuit elements   233
dynamic impedance   134

efficiency   119
electric circuit   11
electric current   1, 12
electric field   25
electricity   10
electromagnetic induction   28
electromotive force (emf)   12
electron   10
energy   3, 11
energy source   11
equivalent circuit   11
    Norton   52
    of transmission lines   216, 221, 223
    Thevenin   48

farad   25
Faraday's Law   28
field   25
    electric   25
    magnetic   28
    vectors   25, 28
filter   129
force   3
form factor   71
frequency   66
    angular   68
    resonant   124
g-parameters   211
gain diagram   131

h-parameters   210
half power frequency   130
henry   29
hertz   66

imagine impedances   226
imaginary axis   84
impedance   78
    characteristic   225
    dynamic   134
    image   226
    of RC circuit   79
    of RL circuit   78
    of RLC circuit   81
    parameters   205
    triangle   78
inductance   28
    coefficient of coupling   31
    energy stored in   35
    in series aiding   32
    in series opposing   33
    mutual   30
    non-linear   29
    self   29
insertion loss   227
integrator
    RL   178
    RC   186
inverse transmission parameters   214
inverse hybrid parameters   211

j-notation   83

Kelvin   2
kilogram   1
Kirchhoff's Laws   40
    Current Law   18, 42
    Voltage Law   16, 42

lag   68
Laplace Transform   192
    application to transient analysis   194
    of an exponential function   193
    of a step function   193
    of the derivative of a function   193
    table of   194
    transform circuits   194
lead   68
length   1
lumped parameters   34

mass 1
magnetic field 28
matrix 141
    addition and subtraction of 142
    cofactor of an element of 145
    column 142
    determinant of 143
    element of 141
    minor of an element of 144
    multiplication of 143
    row 142
Maximum Power Transfer Theorem 54
mesh 41
    current analysis 158–167
metre 1
mutual inductance 30

network 11
neutron 10
nodal voltage analysis 147–157
node 40
non-linear elements
    inductors 29
    resistors 24
Norton's Theorem 52

ohm 14
ohmic material 14
Ohm's Law 13
open circuit 41

partial fractions 196
pass band 130
passive element 11
period 66
periodic time 66
permeability 28
permittivity 25
phase change coefficient 229
phase
    angle 68
    diagram 131
    difference 68
    sequence 108
phasor 70
    diagram 70
phasorial representation of sinusoidal
    quantities 69

π-network 220
polar coordinates 85
potential difference 3, 11
power 3, 36
    apparent 100
    components 100
    dissipated in circuit elements 36
    factor 101
    in balanced three-phase circuits 115
    in single phase circuits 97
    measurement 117
    reactive 100
    real 100
Principle of Superposition 45
propagation coefficient 229
proton 10
pulse 178
    train 187

Q-factor 128
quadrature 84
quantity 1

reactance
    capacitive 76
    inductive 74
rectangular coordinates 85
resistance 13
    dynamic 134
    effect of temperature 22
    internal 22
    non-linear 24
resistors 14
    colour code for 23
    in parallel 18
    in series 16
    non-linear 24
    power dissipated in 36
resistivity 15
resonance 123
    parallel 133
    series 123
resonant frequency 124
root mean square (rms) 70
second 1
short circuit 41
single phase 72

single phase (*cont.*)
  a.c. quantities 72
source
  current 12
  voltage 12
star connection 56, 111
  star-delta transformation 59
steady state 72
step function 172
supermesh 164
supernode 156
susceptance 92
Système internationale d'unites (SI) 1

T-network 222
temperature coefficient of resistance 22
Thevenin's Theorem 48
three phase 107
  balanced three-phase systems 109
  6-wire connection 110
  4-wire connection 110
  3-wire connection 111
time 1
time constant
  of an RL circuit 174
  of an RC network 182
transform
  star-delta 59
  delta-star 56
  Laplace 192
transformer 68
  ideal 223

  practical 224
transients
  double energy 172
  in RC circuits 181
  in RL circuits 172
  in RLC circuits 195, 199
  single energy 172
transmission parameters 213
  lines 216, 221, 223
two port networks 205
  in cascade 219
two wattmeter method 117

units 1
  multiples and submultiples of 6
unit step function 193

vector
  row 142
  column 142
volt 12
voltage 11
  source 12
  division 17

Wattmeter 117
Waveform 66

y-parameters 208

z-parameters 205

Lightning Source UK Ltd.
Milton Keynes UK
UKOW04f1155201216

290465UK00002B/117/P